Additive Manufacturing and Strategic Technologies in Advanced Ceramics

Additive Manufacturing and Strategic Technologies in Advanced Ceramics

Ceramic Transactions, Volume 258

A Collection of Papers Presented at CMCEE-11, June 14–19, 2015, Vancouver, BC, Canada

Edited by
Kiyoshi Shimamura
Soshu Kirihara
Jun Akedo
Tatsuki Ohji
Makio Naito

Volume Editors
Mrityunjay Singh
Tatsuki Ohji
Alexander Michaelis

The American Ceramic Society

WILEY

Published by John Wiley & Sons, Inc., Hoboken, New Jersey.
Published simultaneously in Canada.

Library of Congress Cataloging-in-Publication Data is available.

ISBN: 978-1-119-23600-9
ISSN: 1042-1122

10 9 8 7 6 5 4 3 2 1

Contents

ADVANCED MATERIALS, TECHNOLOGIES, AND DEVICES FOR ELECTRO-OPTICAL AND BIOMEDICAL APPLICATIONS

MULTIFUNCTIONAL COATINGS FOR ENERGY AND ENVIRONMENTAL APPLICATIONS

Preface

The global challenges we face require innovative thinking and sustainable technology to meet increased demands for energy, clean water, and infrastructure. Research in materials, specifically ceramic materials, continues to provide solutions to everyday challenges such as environmental protection, energy supply and generation, and healthcare. The 11th International Symposium on Ceramic Materials and Components for Energy and Environmental Applications (11th CMCEE), held June 14–19, 2015 at the Hyatt Regency Vancouver in Vancouver, B.C., Canada, identified key challenges and opportunities for ceramic technologies to create sustainable materials.

This Ceramic Transactions volume contains papers submitted from the following seven symposia held in Track 4: Cross-Cutting Materials Technologies in Advanced Ceramics:

- Additive Manufacturing Technologies
- Advanced Materials, Technologies, and Devices for Electro-optical and Biomedical Applications
- Multifunctional Coatings for Energy and Environmental Applications
- Novel, Green, and Strategic Processing and Manufacturing Technologies
- Powder Processing Technology for Advanced Ceramics
- Computational Design and Modeling
- Materials for Extreme Environments: Ultra-high Temperature Ceramics (UHTCs) and Nanolaminated Ternary Carbides and Nitrides (MAX Phases)

After a peer-review process, 22 papers were accepted for inclusion in this proceedings volume. The editors wish to extend their gratitude and appreciation to all the symposium co-organizers for their help and support, to all the authors for their cooperation and contributions, to all the participants and session chairs for their time and efforts, and to all the reviewers for their valuable comments and suggestions. We also acknowledge the organization and leadership provided by the meeting chairs, Mrityunjay Singh, Tatsuki Ohji, and Alexander Michaelis.

We hope these proceedings will serve as a useful resource for researchers and engineers working in the field of environmental science and technology.

KIYOSHI SHIMAMURA, NIMS, Japan
SOSHU KIRIHARA, Osaka University, Japan
JUN AKEDO, AIST, Japan
TATSUKI OHJI, AIST, Japan
MAKIO NAITO, Joining and Welding Research Institute, Osaka University, Japan

Additive Manufacturing Technologies

ADDITIVE MANUFACTURING OF MICRO FUNCTIONAL STRUCTURES THROUGH DIAMETER VARIABLE LASER STEREOLITHOGRAPHY AND PRECURSOR SINTERING HEAT TREATMENTS

Soshu Kirihara
Joining and Welding Research Institute, Osaka University
11-1 Mihogaoka Ibaraki 567-0047 Osaka, Japan

ABSTRACT

Stereolithographic additive manufacturing was customized successfully to create micro ceramics components. Photo sensitive acrylic resin with alumina of 170 nm in diameters was spread on a glass substrate with 5 to 10 μm in layer thickness by using a mechanical knife edge. Cross sectional layers patterned by ultraviolet laser scanning of 10 to 100 μm in variable diameter were laminated to create composite precursors. Dense components could be obtained through dewaxing and sintering heat treatments. Photonic crystals with periodic arrangements in magnetic permeability were created to control electromagnetic waves in terahertz frequency range by Bragg diffraction. The terahertz waves can synchronize with vibration modes of various biochemical molecules. Efficient terahertz wave resonators to excite the molecule vibrations will be specifically applied for novel analyzers and reactors. The photonic crystals including micro cavities were designed and fabricated successfully to resonate with microwaves in terahertz frequency range. The wave transmittances through the photonic crystals including the liquid cells were measured by using a terahertz time domain spectroscopy, and cross sectional profiles of electric field intensities were calculated by a transmission line modeling method.

INTRODUCTION

Periodic arrangements in dielectric constants can reflect electromagnetic waves through Bragg diffraction. Especially called photonic crystals theoretically exhibits forbidden gaps prohibiting wave transmissions[1]. The diffraction wavelengths are comparable to the lattice constants. Diamond type micro lattices with isotropic periodicities were processed as the perfect structure to open the photonic band gaps for all crystal directions[2]. However, special lattice propagations were difficult to create by conventional machining of molding processes. In our investigation group, ceramics photonic crystals with diamond structures were created by stereolithographic additive manufacturing and nanoparticles sintering. The band gap formation in the terahertz frequency ranges had been observed[3]. In this investigation, single and double cavities were introduced into the diamond photonic crystals in order to study the characteristic resonance modes by using computer aided design, manufacture and evaluation. These structural misfits can localize the electromagnetic waves strongly through multiple reflections, and wave amplification can realize transmission peak formations in the photonic band gaps according to the defect size[3].

Micro lattices with structural defects can be applied as wavelength filters in the terahertz frequency range. Terahertz waves with micrometer wavelengths and far infrared frequencies are expected to be used in various types of novel sensors to detect dust on electric circuits, defects on material surfaces, cancer cells in human skin, and bacteria in vegetables [4].

EXPERIMENTAL PROCEDURE

Diamond type photonic crystals with or without structural defects were designed using a graphics application (Magics 18, Materialise, Belgium) as shown in Fig. 1. Cavity defects of hollowed unit cells were arranged with center intervals of 1.5 and 2.0, as shown in Figs. 1 (a) and (b), respectively. Dielectric rods of coordination number 4 with an aspect ratio of 1.5 were connected with a coordination number of four to create a diamond structure with a 500 μm lattice constant. The designed model was converted into stereolithography files and sliced into a series of two dimensional layers. These numerical data were transferred to the stereolithographic equipment (SZ-1000S, Sezacc, Japan). Figure 2 shows a schematic illustration of the stereo- lithography fabrication process. Alumina particles (TM-DAR, Taimei Chemicals, Japan) of 170 nm in diameter were dispersed into a photosensitive acrylic resin at 40 vol. % by using a rotation and revolution mixing machine (SK-350T, Shashin Kagaku, Japan). The rotation and revolution speeds were 300 and 700 rpm, respectively. Fluid characteristics of the slurry paste were evaluated by a viscosity and viscoelasticity measuring instrument (VT550, Thermo Scientific, USA). Between two metal disks, the shear stresses were loaded by the rotating upper one on the lower one with a torque meter. The obtained slurry was squeezed onto the working stage from a dispenser nozzle and spread uniformly by a moving knife edge. The squeezing speed was 5 mm/s and the layer thickness was controlled to 10 μm. An ultraviolet laser of 355 nm in wave length is scanned on the ceramics slurry in order to create cross sectional planes with 5 μm in edge part accuracy. The laser beam was valued from 10 to 100 μm in spot size and from 10 to 200 mW in irradiation power. After the formation of the solid pattern, the elevator stage moved downward of 10 μm in the layer thickness, and then the next cross section is stacked. Three dimensional structures are fabricated by stacking all two dimensional layers. A microstructure was obtained through layer stacking of these cross sectional patterns. Uncured resin was removed by ultrasonic cleaning. The composite models were dewaxed at 600 °C for 2 hs and then sintered at 1500 °C for 2 hs in air. The heating rates were 1.0 and 8.0 °C/min in the dewaxing and sintering processes, respectively. The linear shrinkages of the created products were measured by digital optical microscopy (DOM) (VH-Z100, Keyence, Japan). The microstructures of the alumina lattices were observed by scanning electron microscopy (SEM) (JSM-6010LA, JEOL, Japan). A bulk sample of the sintered alumina was also fabricated to measure the relative densities by the Archimedes method. The transmission properties of the photonic crystals, with and without the structural defects, were analyzed to observe the band gap and the localized mode formation by using terahertz time-domain spectroscopy (THz-TDS) (J-Spec, Nippo Precision, Japan). The dielectric constant of the alumina lattice was measured in the sintered bulk sample. A theoretical electromagnetic band diagram of the diamond photonic crystal was graphed through a plane wave expansion (PWE) application (Bandsolve, Rsoft, Japan). Intensity profiles of the structural defects in the presence of electric fields were simulated at localized mode frequencies using a transmission line modeling (TLM) simulator (Microstripes 7.0, Flomerics, UK).

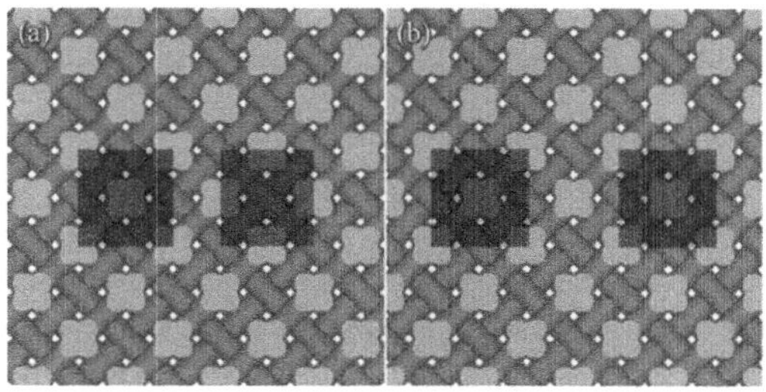

Fig. 1 Graphic models of diamond photonic crystals, with cubic defects indicated by dark areas. Unit cells were hollowed from the lattice structures in order to create double-cavity defects with 1.5 and 2.0 periods in center intervals, as shown in (a) and (b), respectively.

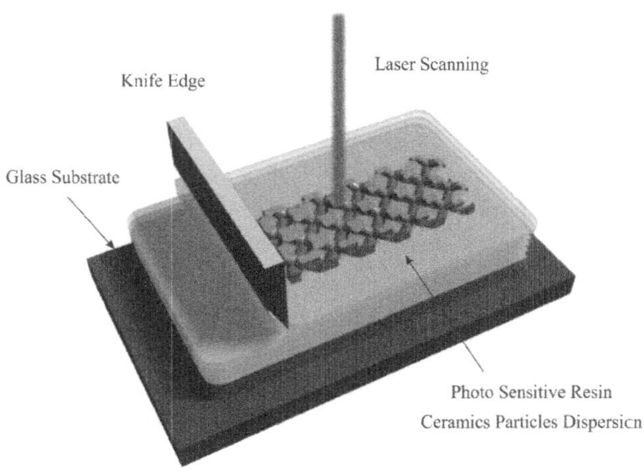

Fig. 2 A Schematic illustration of laser scanning stereolithography of additive manufacturing. Two dimensional layers solidified through laser scanning on photo sensitive resin including with nanoparticles are stacked up to create three dimensional components.

RESULTS AND DISCUSSION

An acryl photonic crystal including with alumina particles fabricated by the micro stereolithography is shown in Fig. 3-(a). The lattice constant of the formed diamond structure was 500 μm. The solid part tolerances can be measured as ±5 μm by DOM. The homogenized dispersions of the alumina particles in the acrylic resin matrix are observed by SEM as shown in Fig. 3-(b). Through the dewaxing and sintering processes, ceramic diamond structures were successfully obtained. Figure 4-(a) shows the sintered diamond structure composed of the micrometer order alumina lattice. The lattice constant and the linear shrinkage were measured by DOM. The lattice constant was measured as 375 μm. The deformation and cracking were not observed. The linear shrinkage on the horizontal axis was 23.8 % and that on the vertical axis was 24.6 %. It was possible to obtain the uniform shrinkage by designing an appropriate elongated structure in the vertical direction for compensation to the gravity effect. The alumina microstructure of 99 % in relative density is observed by SEM as shown in Fig. 4-(b). Dense alumina microstructure was formed, and the average grain size was approximately 2 μm. The forbidden band exhibited in the transmission spectra for the <111>, <100>, and <110> crystal directions were analyzed, and the dielectric constant of the alumina lattice was measured to be 9.0 by using THz-TDS. The higher and lower edges of the gap regions were plotted in the PWE calculated band diagram. The measured results were in good agreement with the calculated results, and a perfect photonic band gap was opened from 0.4 to 0.47 THz. The isotropic propagation of dense alumina lattices with a coordination number of four was verified. These results are evidence that the lattice structures had shrunk equally in all crystal directions without any dimensional deviations during the controlled dewaxing and sintering.

The diamond structure introduced by an air cubic defect with the same dimension as the unit cell is Fig. 5. The transmission spectrum along the Γ-X <100> direction is shown in Fig. 6. Two peaks were observed in the band gap at the frequencies 0.42 and 0.46 THz, respectively. Measured peak frequencies were compared with the simulation by the TLM method as seen in Fig. 7. They were in good agreement with the simulation. The first peak in Fig. 6 was named mode A, while the second one mode B. The electric field distributions of these modes were simulated by the TLM method. Fig. 8 (a) and (b) show cross sectional images of the distributions. In the images, the red area indicates that the electric field intensity is high, whereas blue and green area indicates it is low. Thus, it was considered that the mode A concentrated the oscillation energy of a half wavelength with an antinode in the cube. Also, the mode B concentrated the energy of a half wavelength on the sides of the cube with a node in the cube. Therefore, it was confirmed that the defect introduced structure localized terahertz waves.

The double cavity defects were introduced into the alumina photonic crystals by hollowing the unit cells of the diamond lattices, as shown in Fig. 9. The intensity profiles of electric fields in the vicinity of double cavity defects arranged with 1.5 and 2.0 periods in center intervals, respectively, were simulated by TLM as show in Figs. 10 (a) and (b). The electro-

Fig. 3 Precursor of the photonic crystal fabricated by stereolithography: (a) Acryl rods with a coordination number of 4 and (b) homogenized dispersion of alumina nanoparticles.

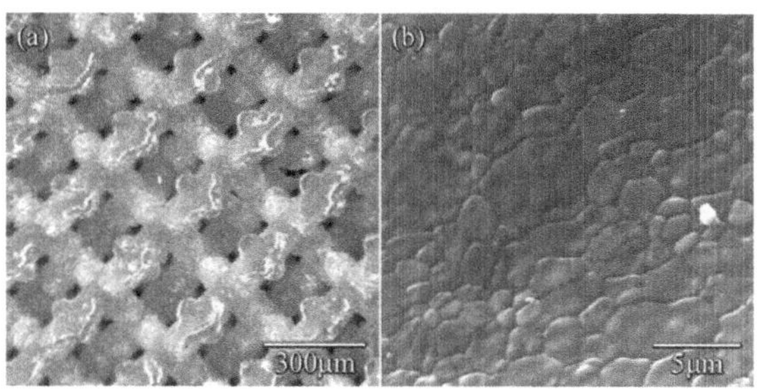

Fig. 4 Full ceramic component of the photonic crystal processed through the dewaxing and sintering. (a) Alumina microlattices with (b) fine microstructure were formed successfully.

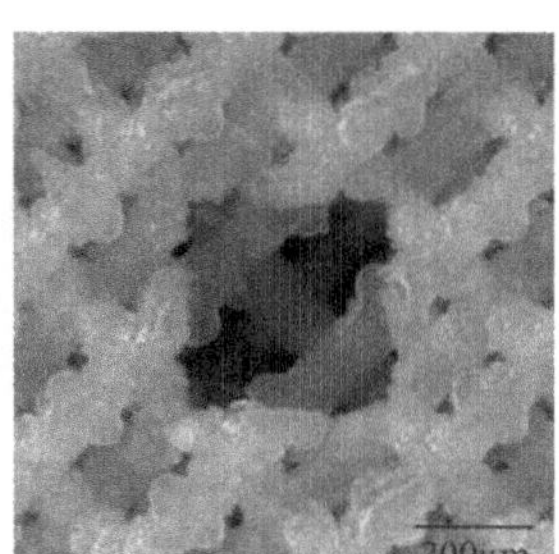

Fig. 5 An alumina lattice with the diamond structure, into which a cubic air defect has been introduced. The edge length of the air cavity is the same size as the lattice constant.

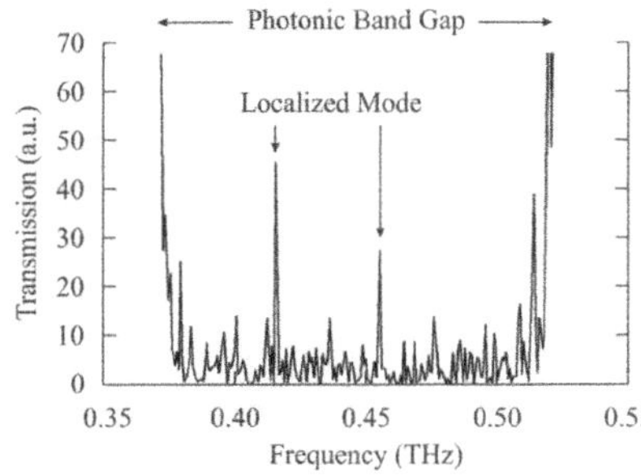

Fig. 6 Transmission spectrum of a structure into which a defect has been introduced. Two peaks representing localized modes of transmission occur in the photonic band gap.

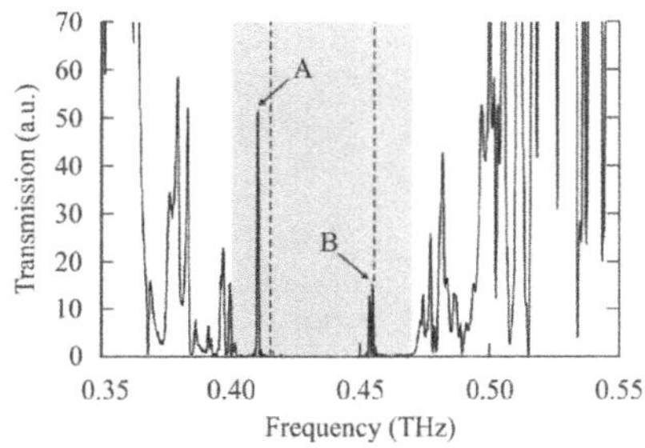

Fig. 7 Simulated spectrum of the structure with the defect. The dashed lines indicate the measured peaks. The frequency range with gray shading indicates the perfect band gap.

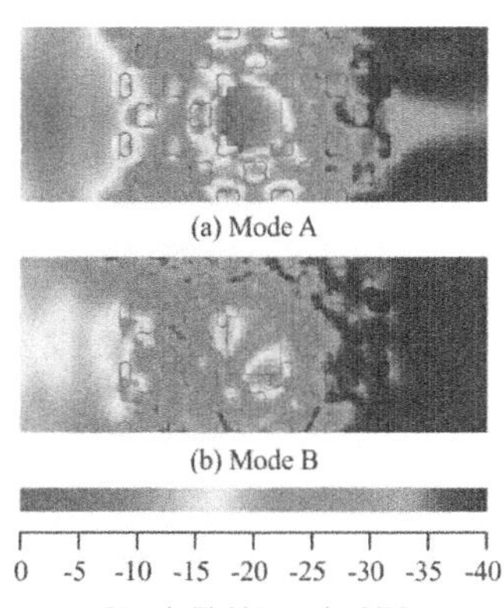

Fig. 8 Electric field distributions of localized mode A (a) and mode B (b) simulated by TLM, using the finite difference time domain (FDTD) technique.

Fig. 9 Sintered alumina lattices with double-defect cavities. The unit cells were hollowed from the diamond structure. The cubic cavities were separated by one period of the lattice .

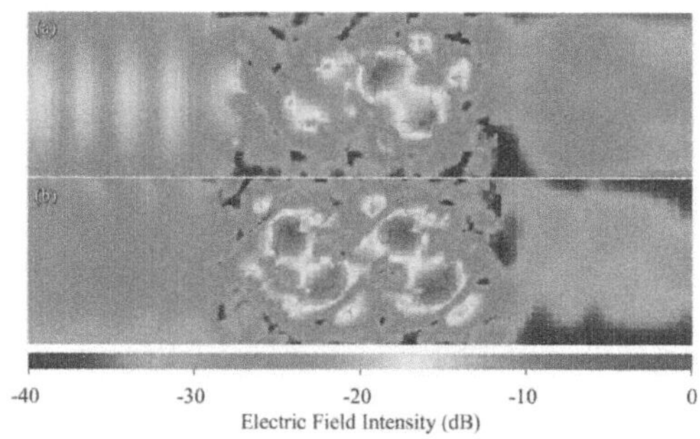

Fig. 10 Intensity profiles of electric fields in the defect cavities. The localized modes were formed in the cavities arranged with 1.5 (a) and 2.0 (b) periods in center intervals.

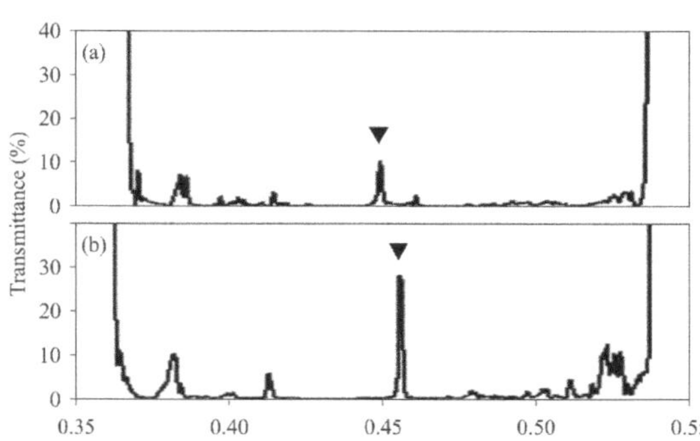

Fig. 11 Transmission amplitude of the terahertz waves through the photonic crystal with defect cavities. The transmission peaks, indicated the solid triangles, were formed thorough localization into the double cavities separated by the 1.5 and 2.0 period, as shown in (a) and (b), respectively.

magnetic waves resonated and localized in the cubic cavities. The single resonance mode spread in the vicinity of closed cavities, and the coupled mode was localized strongly in the separated dual cavity. The transmission spectra, including the photonic band gaps and localized modes, were analyzed by the THz-TDS. As shown in Figs. 11 (a) and (b), the localized mode peaks with transmission intensities of 5 and 27 % formed at 0.45 and 0.56 THz, respectively, through the double-cavity defects separated by the 1.5 and 2.0 periods. The spread mode formed a small resonance peak at the longer wavelength of the lower frequency. The concentrated mode realized sharp and clear peak formation with the higher transmission intensity.

CONCLUSIONS

Through stereolithographic additive manufacturing and nanoparticles sintering, alumina photonic crystals with diamond structures were successfully fabricated in order to control terahertz waves through Bragg diffraction. Acryl models with alumina nanoparticles were dewaxed and sintered carefully and fine diffraction lattices were obtained according to the designed models. By the introduction of single and double cavities could, localized modes of transmission peaks could be created in complete band gaps. Filtering properties at specific wavelengths can be applied in sensor devices.

REFERENCES

[1]K. Ohtaka, Energy Band of Photons and Low-energy Photon Diffraction, Physical Review B, 19 [10] (1979) 5057-5067.

[2]K. M. Ho, C. T. Chan and C. M. Soukoulis, Existence of a Photonic Gap in Periodic Dielectric Structures, Physical Review Letters, 65 [25] (1990) 3152-3155.

[3]S. Kirihara, Y. Miyamoto, International Journal of Applied Ceramic Technology, 6 (2009) 41.

[4]B. Temelkuran, Mehmet Bayindir, E. Ozbay, R. Biswas, M. M. Sigalas, G. Tuttle and K. M. Ho, Photonic crystal-based Resonant Antenna with a Very High Directivity, Journal of Applied Physics, 87 [1] (2000), 603-605.

[5]H. Yada, M. Nagai, and K. Tanaka, The Intermo-lecular Stretching Vibration Mode in Water Isotopes Investigated with Broadband Terahertz Time-domain Spectroscopy, Chemical Physics Letters, 473 [4-6] (2009) 279–283.

STEREOLITHOGRAPHIC ADDITIVE MANUFACTURING OF SOLID ELECTROLYTE DENDRITES WITH ORDERED POROUS STRUCTURES FOR FUEL CELL MINIATURIZATIONS

Soshu Kirihara
Joining and Welding Research Institute, Osaka University
11-1 Mihogaoka Ibaraki 567-0047 Osaka, Japan

ABSTRACT

Solid electrolyte dendrites of yttria stabilized zirconia with spatially ordered porous structures were successfully fabricated for fuel cell miniaturizations by stereolithography of additive manufacturing. Micrometer order ceramic lattices with the coordination numbers 4, 6, 8 and 12 were propagated spatially in computer graphic space. The lattice aspect ratios were designed between 1.0 and 2.0 to value the porosities in higher levels from 50 to 80 %. Nanometer sized yttria stabilized zirconia were dispersed into photo sensitive liquid resins at 30 % in volume fraction. The paste material was spread on a glass substrate with 10 μm in layer thickness by mechanic knife edge movements, and an ultra violet micro pattern was exposed on the surface to create cross sectional solid layer with 2 μm in part accuracy. After the layer stacking process, the ceramic dispersed resin lattices of 100 μm in diameter were obtained exactly. These composite precursors were dewaxed and sintered at 600 and 1500 °C in an air atmosphere, respectively, and the fine ceramic lattices of 98 % in relative density were created. Gaseous fluid profiles and pressure distributions in the formed ceramic lattices with the various coordination numbers and porosity percent were visualized and analyzed by using finite difference time domain simulations.

INTRODUCTION

For sustainable development and emission reduction of carbon oxides, solid oxide fuel cells (SOFCs) are actively investigated as novel generation systems of electric powers with high efficiencies in energy conversion circulations. Yttria stabilized zirconia (YSZ) with higher ion conductivities for incident oxygen is widely adopted material for solid electrolyte anodes as the SOFC components[1-5]. To increase surface areas of reaction interfaces and gap volumes of stream paths, porous network structures have been introduced into YSZ electrodes in micrometer or nanometer sizes[6-8]. In our investigation group, micrometer order ceramics components of photonic crystal with dielectric lattices and artificial bones with biological scaffolds were fabricated successfully by stereolithographic additive manufacturing[9-11]. Micrometer order porous structures were verified to be able to modulate energy and material flows effectively according to geometrical designs and theoretical simulations. In this investigation, solid electrolyte dendrites composed of YSZ spatial lattice structures with various coordination numbers were fabricated successfully by using micro pattering stereolithography and powder sintering techniques. In the dendrite structure, stress distributions and fluid flows were simulated and visualized by using finite element methods. In geometrical lattice modulations, coordination numbers and aspect ratios were optimized effectively to increase the reaction surfaces and realize the smooth gaseous propagations. In our research group, computer aided designs, manufactures and evaluations have been established to create micro ceramics components.

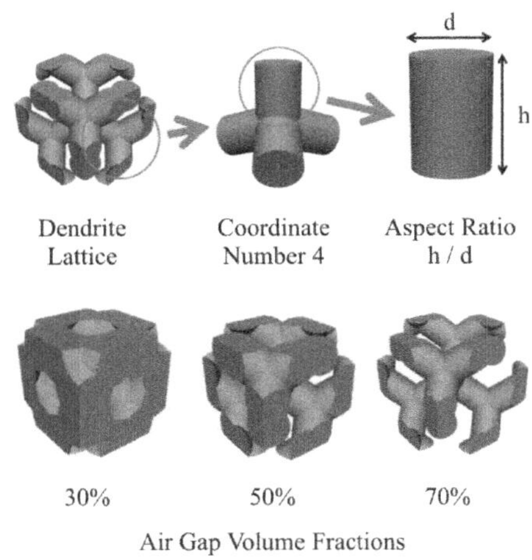

Fig. 1 Computer graphics of lattice distributions of dendrite structures with coordination number four. Air gap volume fractions are changed with aspect ratios.

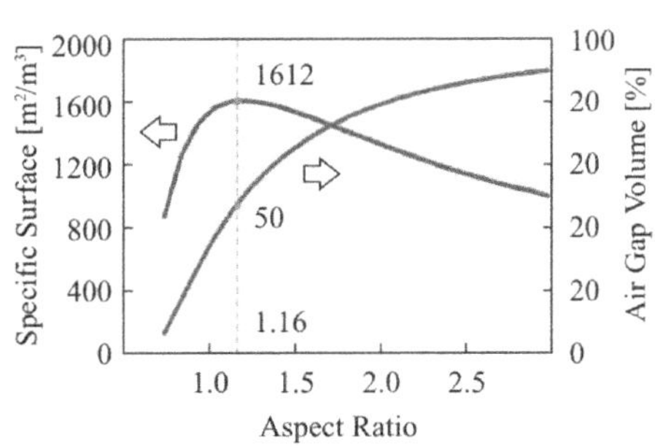

Fig. 2 Optimization of the aspect ratio to obtain the widest specific surface and higher air gap volume. The aspect ratio was decided as 1.16 for designing.

GEOMETRICAL DESIGN

The spatial dendrite structures of solid electrolyte were designed by using a computer graphic application (Think Design: T3, Japan) as shown in Fig 1 . The coordination number was decided as 4 to increase the surface area and realize higher mechanical properties of the lattice module. The ordered dendrite structures can realize larger surface areas and higher gap volumes comparing with random porous structures. The surface areas of reaction interfaces and the gap volume of stream paths were calculated geometrically for the dendrite lattice with coordination number 4 as shown in Fig 2. The dendrite lattices of 1.16 in aspect ratio can be considered to exhibit the higher reaction efficiencies and gas transmittances according to Nernst equation. In the optimized dendrite structure, the diameter and length of the YSZ rods were decided 92 and 107 μm, respectively. The ideal lattice constant was estimated 250 μm. The graphic data was converted into a stereolithographic format through polyhedral approximations. The solid model was sliced into the cross sectional numerical data sets by using graphic application (Materialize: Magics, Belgium) to input the stereolithographic equipment (D-MEC: SI-C1000, Japan,)

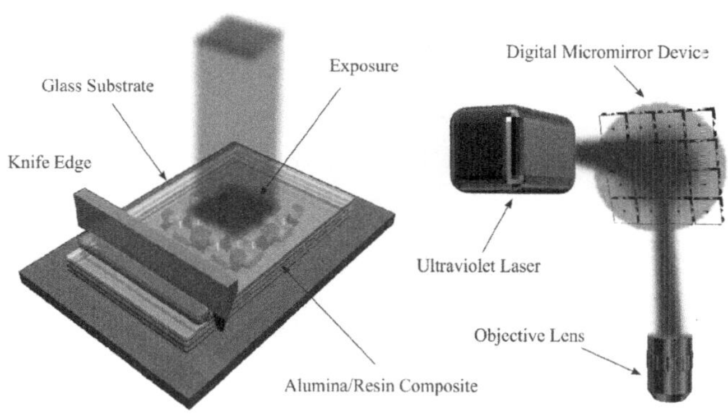

Fig. 3 A schematic illustration of micro patterning stereolithography. Fine images are exposed by using a digital micro mirror device on a photo sensitive resin.

EXPERIMENTAL PROCEDURE

Photo sensitive acrylic resin dispersed with YSZ particles of 60 and 100 nm in first and second diameters at 30 volume % were fed over a substrate from a dispenser nozzle. The highly viscous resin paste was fed with controlled air pressure, and spread uniformly by a mechanical knife edge. The thickness of each layer was controlled to 10 μm. The cross sectional pattern was formed through illuminating visible laser of 405 nm in wavelength on the resin surface. Figure 3 shows a schematic illustration of the micro patterning stereolithography. The high resolution image could be achieved by applying a digital micro mirror device (DMD) and an objective optical lens.

In this optical device, micro aluminum mirrors of 14 μm in edge length were assembled with 1024×768 in numbers. Each mirror can be tilted independently by piezoelectric actuating. The ultra violet lay of 405 nm is introduced into the digital micro mirror device, and the cross sectional image is reduced at 1/10 through an objective lens set and concentrated into the exposing area 1.3×1.7 mm in size. Through the layer stacking under the computer control, the acrylic resin component with the ceramics particles dispersion is obtained. In order to avoid deformation and cracking during dewaxing, careful investigation for the heat treatment processes were required. The formed precursors with dendrite structures were heated at various temperatures from 100 to 600 °C while the heating rate was 1.0 °C /min. The dewaxing process was observed in respect to the weight and color changes. The YSZ particles could be sintered at 1500 °C for 2 hs. The heating rate was 8.0 °C/min. The density of the sintered sample was measured by using Archimedes method. The ceramic microstructures were observed by a digital optical microscope and scanning electron microscopy. In the lattice dendrites, fluid flow velocities and pressure stress distributions were simulated and visualized by a finite volume method (FVM) application (Ansys: Cybernet Systems, Japan).

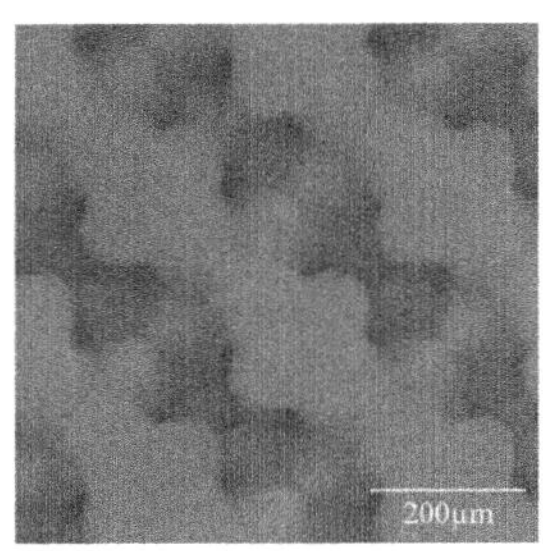

Fig. 4 Acryl dendrite lattice with YSZ particles fabricated by using the micro stereolithography.

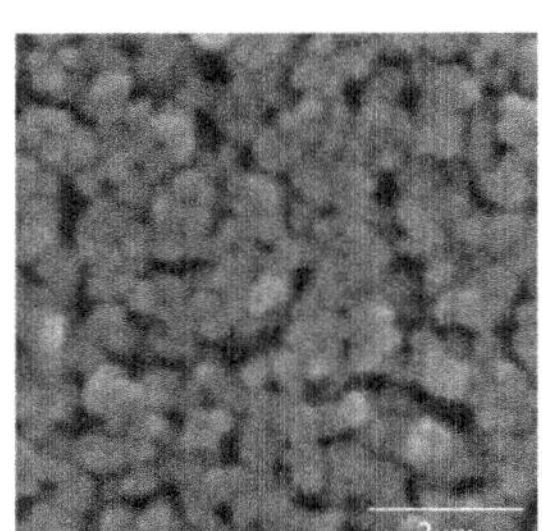

Fig. 5 The YSZ nanoparticles dispersed in the acryl micro lattice. The volume fraction is 30 %.

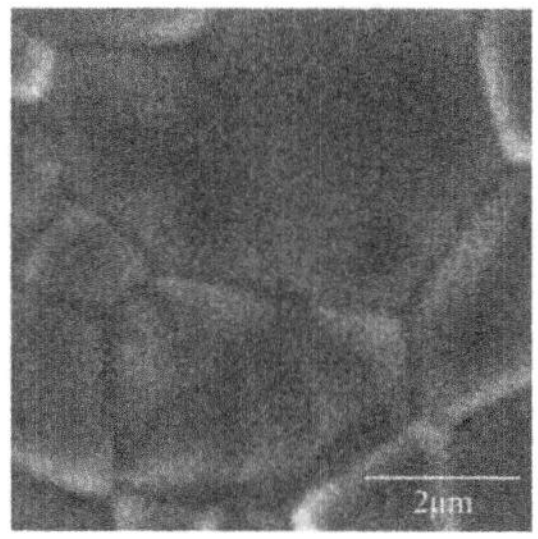

Fig. 6 The microstructure of YSZ dendrite lattice. The average grain size was measured as 4 μm.

Fig. 7 The sintered dendrite lattices of YSZ solid electrolyte. The lattice part accuracy is 2 μm.

RESULTS AND DISCUSSION

The acryl dendrites including with the YSZ particles were processed exactly by the micro pattering stereolithography as shown in Fig. 4. The spatial resolution in the model was approximately 1.0 %. The microstructure of the composite lattice is shown in Fig. 5. The nanometer sized YSZ particles were dispersed homogeneously in the acryl matrix. The formed precursors with dendrite structures were heated at various temperatures from 100 to 600 °C while the heating rate was 1.0 °C /min. The dewaxing process was observed in respect to the weight and color changes. The YSZ particles could be sintered at 1500 °C for 2 hs. The heating rate was 8.0 °C/min. The linear shrinkages on the both horizontal and vertical axis were 32 %. In the dense microstructure of the YSZ lattice as shown in Fig. 6, the average grain size was approximately 4 μm. The density of the sintered sample was measured by using Archimedes method. The relative density reached at 95 %. Micrometer sized cracks or pores were not observed. Figure 7 shows the sintered solid electrolyte dendrite with the YSZ micro lattice structure. The macroscopic deformations were not observed. The volume fraction of the air gaps was 50 % by the open paths. In the other previous investigations, the porous electrodes were formed by sintering the YSZ surly with polystyrene particles dispersion. Therefore, it is difficult to realize the prefect opened pores structures with the higher porosity over 40 % in volume fraction. Micrometer sized cracks or pores were not observed. The obtained dense YSZ lattice structure will be exhibit the higher performances in mechanical properties as the porous electrodes of the solid electrolyte dendrites. The fluid flow velocities are visualized by using the FVM method as shown in Fig. 8. All air paths are opened for outsides and connected with each other in the YSZ dendrite lattice structures. The fluid flows can transmit the one direction smoothly. The pressure stress distributions in the dendrite are visualized as shown in Fig. 9. The fluid pressures are gradually distributed for flow direction, and the localization of the stress is not observed. The fabricated solid electrolyte dendrites with YSZ lattices can be considered to have higher performances as novel ceramic electrodes. The fabricated solid electrolyte dendrites with YSZ lattice structures can be considered to have higher performances as novel ceramic electrodes in near future SOFCs.

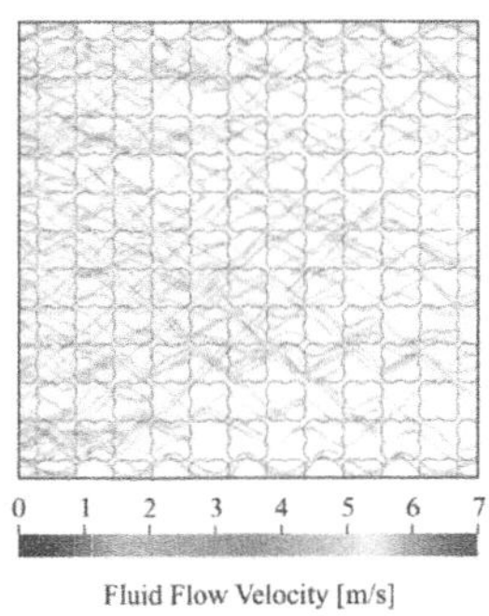

Fig. 8 A distribution of fluid flow velocities in the dendrite lattice structure simulated and visualized by FVM. The curved lines show the fluid flow paths according to the velocity vectors.

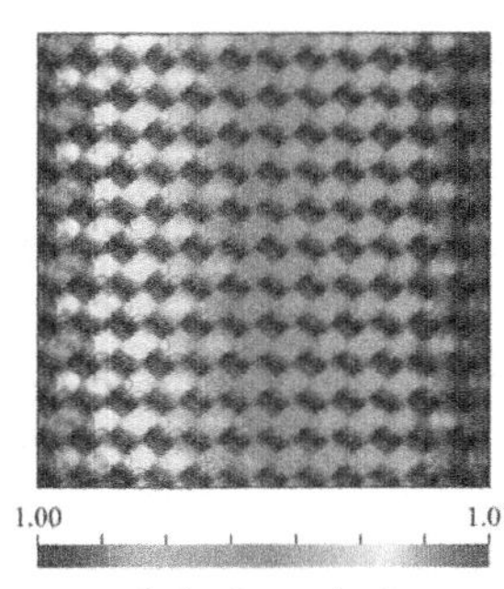

Fig. 9 The distribution of the surface pressure on the ceramics dendrite lattices. The red and blue areas show the higher and lower gas pressures on the reaction interfaces, respectively.

CONCLUSIONS

Solid electrolyte dendrites composed of yttria stabilized zirconia for anode electrodes of solid oxide fuel cells were fabricated. Acryl precursors including ceramic particles were formed successfully by stereolithographic additive manufacturing. Thorough careful optimization of process parameters in dewaxing and sintering, dense ceramic components were obtained. Opened path networks in porous electrolytes exhibited effective transmission properties of gaseous flows. The ceramics stereolithography will contribute to fuel cells miniaturizations.

REFERENCES

[1]N. Q. Minh, "Ceramic Fuel Cells", Journal of American Ceramic Society, Vol. 76, No. 3, pp. 563–588, March, 1993.

[2]S. C. Singhal, "Solid oxide fuel cells for stationary, mobile, and military applications", Solid State Ionics, vol. 152-153, pp. 405-410, December, 2002.

[3]A. B. Stambouli, E. Traversa, "Solid Oxide Fuel Cells: a Review of an Environmentally Clean and Efficient Source of Energy", Renewable Sustainable Energy Reviews, vol. 6, No. 5, pp. 433–455, October, 2002.

[4]T. A. Ramanarayanan, S. C. Singhal, E. D. Wachsman, "High Temperature Ion Conducting Ceramics", The Electrochemical Society Interface, Vol. 10, No. 2, pp. 22-27, Summer, 2001.

[5]J. Will, A. Mitterdorfer, C. Kleinlogel, D. Perednis, L. J. Gauckler, "Fabrication of thin electrolytes for second-generation solid oxide fuel cells", Solid State Ionics, Vol. 131, No. 1-2, pp. 79-96, June, 2000.

[6]J. Hua, Z. Lüa, K. Chena, X. Huanga, N. Aia, X. Dub, C. Fub, J. Wanga, W. Su, "Effect of composite pore-former on the fabrication and performance of anode-supported membranes for SOFCs", Journal of Membrane Science, Vol. 318, No. 1-2, pp. 445-451, June, 2008.

[7]J. J. Haslam, A. Q. Pham, B. W. Chung, J. F. Dicarlo, R. S. Glass, "Effects of the use of pore former on performance of an anode supported solid oxide fuel cell", Journal of American Ceramic Society, Vol. 88, No. 5, pp. 513-518, March, 2005.

[8]T. Talebi, M. H. Sarrafi, M. Haji, B. Raissi, A. Maghsoudipour, "Investigation on microstructures of NiO-YSZ composite and Ni-YSZ cermet for SOFCs", International Journal of Hydrogen Energy, Vol. 35, No. 17, pp. 9440-9447, September, 2010.

[9]S. Kirihara, Y. Miyamoto: "Terahertz Wave Control Using Ceramic Photonic Crystals with Diamond Structure Including Plane Defects Fabricated by Micro-stereolithography", The International Journal of Applied Ceramic Technology, Vol. 6, No. 1, pp. 41-44, January, 2009.

[10]S. Kirihara, T. Niki, M. Kaneko, "Terahertz Wave Behaviors in Ceramic and Metal Structures Fabricated by Spatial Joining of Micro-stereolithography, Journal of Physics, in printing.

[11]S. Kirihara, T. Niki, M. Kaneko, "Three-dimensional Material Tectonics for Electromagnetic Wave Control by Using Micoro-stereolithography", Ferroelectrics, in printing.

Processing of thermoplastic suspensions for Additive Manufacturing of Ceramic- and Metal-Ceramic-Composites by Thermoplastic 3D-Printing (T3DP)

U. Scheithauer, E. Schwarzer, A. Haertel, H.J. Richter, T. Moritz, and A. Michaelis
Fraunhofer Institute for Ceramic Technologies and Systems, Dresden, Germany

ABSTRACT

Our new approach in the field of Additive Manufacturing (AM) – thermoplastic 3D printing (T3DP) – is presented. High-filled ceramic and metal feedstocks, based on thermoplastic binder systems, were used for AM of dense ceramic components and metal-ceramic-composites. Compared to other AM methods, the developed T3DP method has outstanding advantages. Two benefits can be highlighted. The first one concerns the portfolio of applicable materials, which is not limited. The second one refers to the process, which works nearly independent from the physical properties of the dispersed powder material.

The application of T3DP in principle was demonstrated by samples, made of alumina and zirconia as well as stainless steel-zirconia composites. Different feedstocks with high powder contents up to 67 vol.-% were prepared. For the sintered samples a density of about 99 % and higher was obtained. FESEM studies of the samples' cross-section presented a homogenous microstructure.

With regard to steel-zirconia composites, the adjustment of the sintering shrinkage behavior was the main challenge. The sinterability of the metal powder could be increased by high energy milling and an adjustment of the powder content.

The adapted AM method of T3DP offers new prospects for fabrication of multi-material components. This AM method will be applicably not only for steel-zirconia composites, but also for any other combination of materials which can be processed to a paraffin based thermoplastic feedstock and which can be co-sintered.

INTRODUCTION

Multi-material components or components with a graded microstructure open new fields of applications or allow the miniaturization of components. For instance by the combination of steel and ceramics, innovative multi-functional properties in two- and in three-dimensions can be achieved, such as hard and ductile, electrically or thermally conductive and insulating, magnetic and nonmagnetic. Possible applications are in a variety of industrial and medical fields, for example as cutting tools, wear resistant components, energy and fuel cell components or as bipolar surgical tools[1-4].

However different restrictions concerning the possible material combinations have to be noted. The thermal expansion coefficient (TEC) has to be conformable to avoid mechanical stresses which were initiated by a change of the temperature, e.g. during cooling after the sintering process.

Furthermore processible powders or suspensions have to be prepared made by the chosen materials. The debinding and sintering behavior of the selected powders have to be adjusted by modification (particle size distribution, particle shape) and the solid content within the feedstocks/suspensions to avoid crack initiation during the thermal processing.

Differences in the thermal expansion or in the shrinking behavior of the materials can induce high stresses which are the reason for defects like structural changes, warpage delaminations or cracks[4, 5].

Additive manufacturing (AM) comprises manufacturing processes where objects are built up layer-by layer. According to ASTM additive manufacturing is a "process of joining material to make objects from 3D model data, usually layer upon layer"[6]. AM technologies are tool-free production processes which increase the freedom for the component design and allow designing and constructing oriented to function instead to fabrication.

AM of polymers is state-of the-art and in the field of metals, processing of more and more materials almost accomplished. For processing ceramic materials, the technical application of AM technologies has been limited so far.

Our main goal is the development of an AM technology, which can be used for the production of multi-material-components to combine multi-functional properties with freedom in design and construction. A technology is needed which meets following requirements:

- processing of ceramic or/and metallic materials
- selective deposition of different materials in different layers as well as in one layer (Figure 1)
- precise deposition of small volumes and fast deposition of large volumes
- solidification independently from material properties
- removable binder system at different atmospheres (oxidizing, reducing, inert) to avoid any material conversion

Furthermore, a combination of different dispensing technologies, on the one hand side for the precise deposition of small volumes (e.g. a pulsing micro-droplet system based on piezo technique) and on the other side fast deposition of large volumes (e.g. with a type of dispensing technique for filaments including a changeable nozzle technique for different filament diameters) is preferred to increase the productivity and resolution (Figure 1).

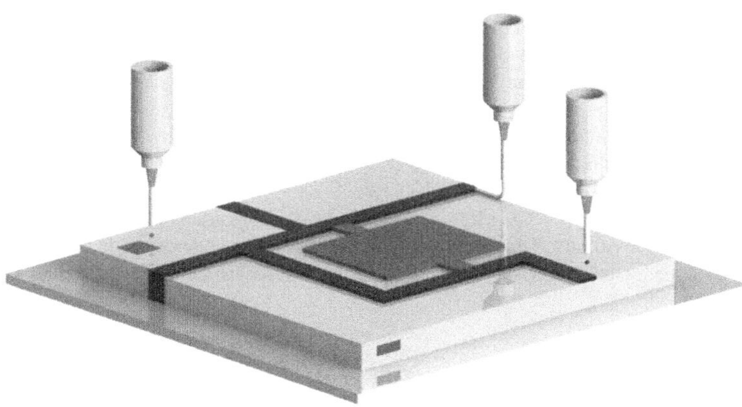

Figure 1. Schema for AM of multi-material components and combination of different dispensing technologies for suspensions.

However, ceramic materials have been studied in AM processes ab initio with the development of the different AM technologies since about 25 years, see e.g.[7, 8]. AM technologies can be classified according to the state of the material type that is used[9] – powder materials, liquid materials and solid materials. It is also possible to classify the AM technologies according the dimensional order[10] – point, line or plane.

Conventional stereolithography (STL) for example, has been applied to alumina, zirconia, silicon nitride and silica as well as to ZTA-ceramics[11-14]. In this STL-process a light-curable ceramic slurry is cured by an UV-laser. Binders that are cured under visible blue light were applied in a DLP (direct light processing) process, which allows the production of complex shaped dense alumina parts[15]. The DLP process has a high potential for producing monolithic dense ceramics. However, the production of composite materials is impossible with customary DLP equipment, caused by the limitation of only one suspension for the manufacturing of the whole structure. Despite this, in DLP the material variety is limited. The reason is, that the powder must not absorb light of the wavelength that initiate the photo-polymerization of the binder. Only white or bright powders can be processed. The light absorption of dark powders, e.g. metal powders, prevent the penetration of the light radiation even in thin layers and the photo-polymerization cannot start.

Selective laser sintering (SLS), an AM technology which is established for some metals in industrial processes has been tested for a number of ceramic materials, too[16-20]. But it is difficult to generate a dense microstructure in ceramic components without any glassy phase or without significant grain growth.

Another technique, the 3D powder bed printing, is the most commonly used AM process for powder materials. Typical applications of 3D powder bed printing focus on the production of porous ceramic components because of the lowly compacted powder layers; the green density is too low to reach high density after sintering. Furthermore it is challenging to apply different materials in one layer. An interesting approach was presented by Polsakiewicz[21]. Metallic and ceramic inks were printed in a powder bed to generate multi-material components. But only small particles can be printed and placed between the great particles of the powder bed.

In general, when using AM methods for the production of materials with high sinter density, it is necessary to use a suspension, paste or feedstock with a high powder volume content instead of a dry powder bed. This was shown for example in SLS where instead of powder layers suspension layers were deposited and consolidated (after short drying) by laser beam[22].

Furthermore, the direct printing of suspensions has been shown as a promising way to produce dense ceramic materials by additive manufacturing[23, 24]. Of course, only highly dispersed nano- or submicron-powders can be processed by direct printing using customary print heads with very small diameters.

Conventional fused deposition modeling (FDM) has been tested for example to produce functional ceramic materials[25] and alumina[26]. However, the effort for the preparation of the thermoplastic ceramic feedstock in form of spooled filaments limits the FDM application for ceramics as well as the high viscosity of the molten suspensions and the high wear, which increase the requirements on the dispensing aggregates.

The robocasting process is a computer-controlled deposition of colloidal pastes or suspensions. In contrast to original FDM, the carrier fluid is a volatile solvent (water or organic liquid). In robocasting highly dispersed ceramic suspensions are used for AM of complex ceramic structures[27, 28].

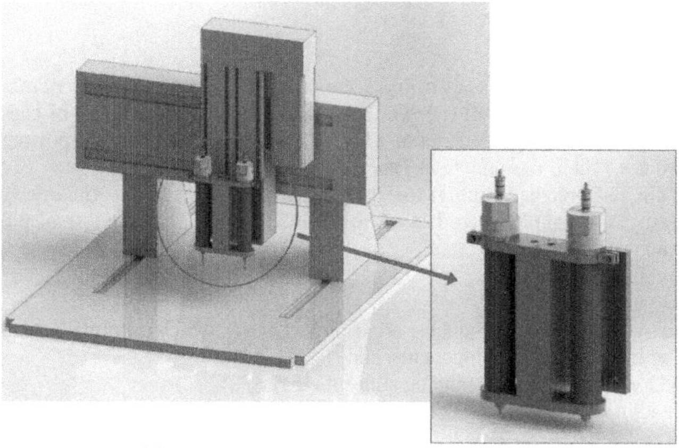

Figure 2. Heatable dispensing units with xyz positioning as laboratory equipment (schema).

In our new approach, which is called thermoplastic 3D printing (T3DP), molten, thermoplastic feedstocks are used which are handled in a dispensing unit with xyz positioning (Figure 2). The feedstocks base on compositions known from low-pressure injection molding[29, 30]. The melting temperature is relatively low (approx. 100 °C) and the viscosity too (< 10 Pas @ 100 s-1) compared to thermoplastic feedstocks used in conventional FDM. We used a molten feedstock which can be simply dispensed via a thin nozzle due to its relatively low viscosity. This T3DP-concept has several advantages. On the one hand, there are no restrictions concerning the applied powder material because the consolidation occurs by increasing of viscosity during cooling. On the other hand, composite materials can be produced by using two or more dispensing heads. Basically, it is possible to process all materials which can be dispersed in the molten feedstock.

Former works showed the suitability of the T3DP for the AM of nearly dense alumina and zirconia components[31], but also metal powders could be processed[32] as well as metal-ceramic-composites[33]. Table 1 shows the materials which were processed with T3DP.

Table 1. Used materials and properties for single material samples made by T3DP.

material	type, company	d_{50}	Vol.-%	$T_{sintering}$ (atmosphere)	porosity (%)
alumina	MR52, Martinswerk	1.4 µm	67	1600 °C (air)	0.8
zirconia	TZ-3Y-SE, Tosoh	0.2 µm	45	1500 °C (air)	0.1
	TZ-3Y-E, Tosoh	0.1 µm	50	1400 °C (air)	0.1
stainless steel	17-4PH, Sandwik	12.2 µm	47	1365 °C (Ar/H2)	<1.0
	CroFer22APU, ThyssenKrupp	24.6 µm	60	1365 °C (Ar/H2)	25.0
				1400 °C (Ar/H2)	13.0
	CroFer_HEM, IKTS	9.4 µm	50	1400 °C (Ar/H2)	3.0

Zirconia and stainless steel are well suited for metal-ceramic-composites, which is due to the fact that both materials have almost similar TEC, sintering temperature and atmosphere, which has been verified for example in[34-36]. But sintering had to be adjusted which was realized by a powder modification by high energy milling presented in a former paper[33].

One goal of this work was to investigate the behavior of different suspensions during the thermal debinding under different atmospheres. Another aim was to investigate the possibility to form very small droplets and to deposit these droplets on a substrate.

EXPERIMENTAL PROCEDURES

Debinding behavior

To characterize the thermal degradation, suspensions were produced from three different materials (alumina, zirconia, and stainless steel (17-4PH)). The suspensions based on a mixture of bees wax and paraffin and were produced and homogenized in a heatable ball mill. The realized powder content corresponded to the values, which were quote in Table 1.

STA 409 C/CD (NETZSCH-Gerätebau GmbH) was used for simultaneous thermo gravimetry and differential thermal analysis. The samples were heated up to 740 °C with a heating rate of 0.6 K/min. Four different atmospheres were used:

- synthetic air (air) as oxidizing atmosphere,
- argon (Ar) as inert atmosphere,
- a mixture of argon and hydrogen (4:1) as reducing atmosphere (ArH_2), which is normally used for sintering of metal or metal-ceramic-components, and
- a mixture of argon and hydrogen (1:9) as reducing atmosphere (H_2)

and the volume flow rate was adjusted to 5 l/h.

Deposition of small droplets

The possibility to deposit the T3DP-suspensions as filament was demonstrated in former works. This process can be used to manufacture large structures in a short time caused by the high volume rate within the heatable dispensing unit and the used needles with inner diameters between 0.4 and 0.8 mm. Unfortunately, dispensing of very small volumes failed, which would be necessary to manufacture thin structures.

The heatable micro dispensing system MDV 3200A-HS (Vermes, Germany), a system for single drop deposition, was tested with a ceramic slurry composition. This system consists of a piezo actor which is actuated by a control unit, a tappet inside a small chamber, which opens and closes a nozzle, and lever system connecting the piezo actor with the tappet.

RESULTS

Debinding behavior

Figures 3 to 6 show the results of the different thermal degradation tests. The mass of the binder components is plotted as function of the temperature. All curves show a similar trend. Up to a temperature of about 150 °C no change in the mass was detected. Afterwards the mass decreases rapidly with a maximum negative slope at a temperature of about 200 °C. At higher temperatures the curves show a different behavior for the three materials but nearly a similar behavior for the four different atmospheres.

The tests showed that the debinding behavior of the tested suspension is a function of the dispersed material, but not of the used debinding atmosphere. But for all tests it was possible to remove the binder system completely by thermal degradation.

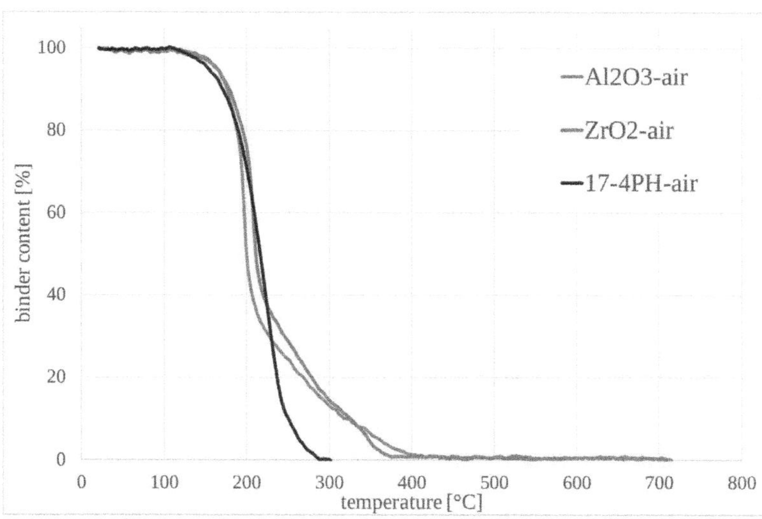

Figure 3. Binder content vs. temperature for debinding under oxidizing atmosphere.

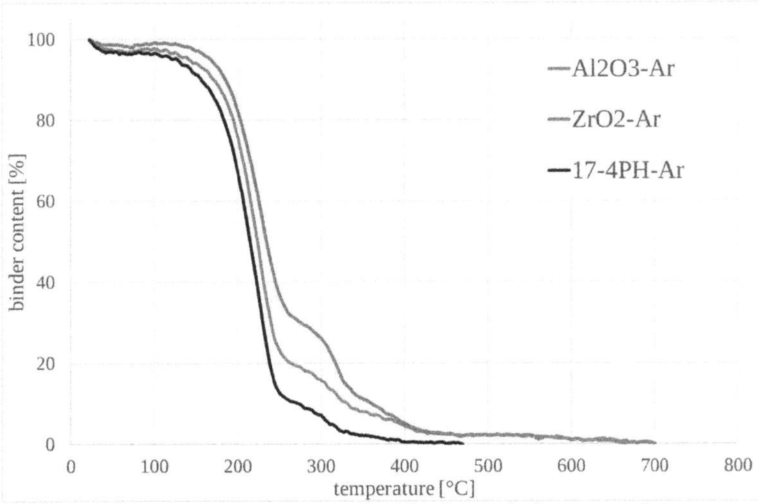

Figure 4. Binder content vs. temperature for debinding under inert atmosphere.

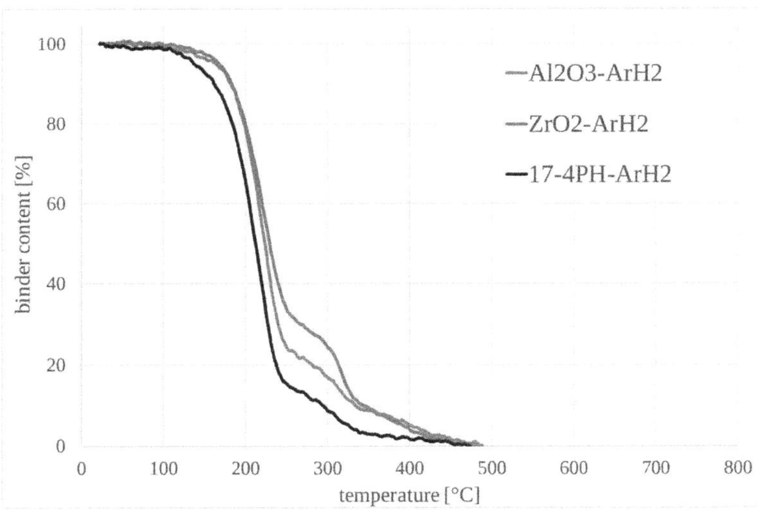

Figure 5. Binder content vs. temperature for debinding under reducing atmosphere ($Ar:H_2$=1:9).

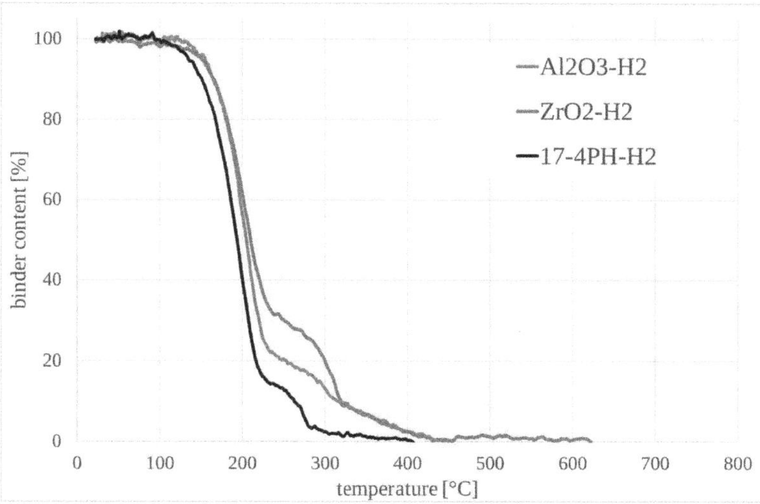

Figure 6. Binder content vs. temperature for debinding under reducing atmosphere (Ar:H_2=4:1).

Deposition of small droplets

First tests with the heatable micro dispensing system were carried out with the alumina suspension, which had a powder content of about 67 vol.-%. During these experiments it could be noticed that it is impossible to carry the suspension from a heatable reservoir in the heated chamber only by applying pressure on the top of the suspension reservoir. Therefore it was necessary to decrease the viscosity, which was realized by decreasing the powder content down to 64 vol.-%. The flow behavior of the suspensions in dependence of shear rate and temperature was characterized by rheological measurements using a rheological measurement device (Rheometer MCR 302) from Anton Paar. Therefore, a plate-plate device (diameter of 25 mm) was used. The parameters for the measurements were an increasing shear rate from 0.01 to 100 s^{-1} for different constant temperatures during the measurement (70 to 110 °C). Every sample was heated in the system for 20 minutes to the selected temperature, achieving homogeneous heat distribution, before starting the measurement. Figure 7 shows the rheological behavior of the two different suspensions.

Both suspensions showed a shear thinning behavior. The viscosity decreases with increasing shear rates significantly, as well as for low shear rates and increasing temperature. This behavior allows the selective deposition of the suspensions which stay at the selected place because of the increasing viscosity after deposition resulting from the cooling and the missing shear stresses.

Commonly, the viscosity decrease with an increasing temperature and for suspension with a decreasing particle content. Figure 7 shows a behavior conform to the theory, but for small shear rates, the trend differs. Here, the suspensions with a particle content of 64 vol.-% show indeed a decreasing viscosity with an increasing temperature, but higher compared to the suspensions with a particle content of 67 vol.-%. A possible explanation could be the slippage of the suspension caused of the higher amount of binder, especially beeswax and paraffin. These components have lubricating properties, which normally increase at higher temperatures. This phenomenon can be seen and gets more significant caused of the lower particle content. Based on this, the rheological measurement containing errors at small shear rates, in the range of nearly 2 to 30 s^{-1}, caused of lubrication. Another indication for this case is given in Figure 8, which shows the shear stress versus shear rate at 70 and 100 °C for the different suspensions.

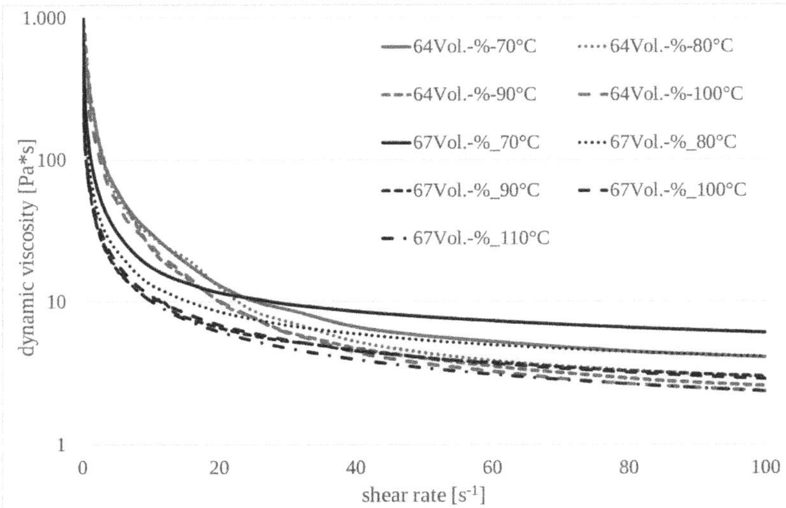

Figure 7. Dynamic viscosity vs. shear rate for two Al_2O_3-suspensions at different temperatures.

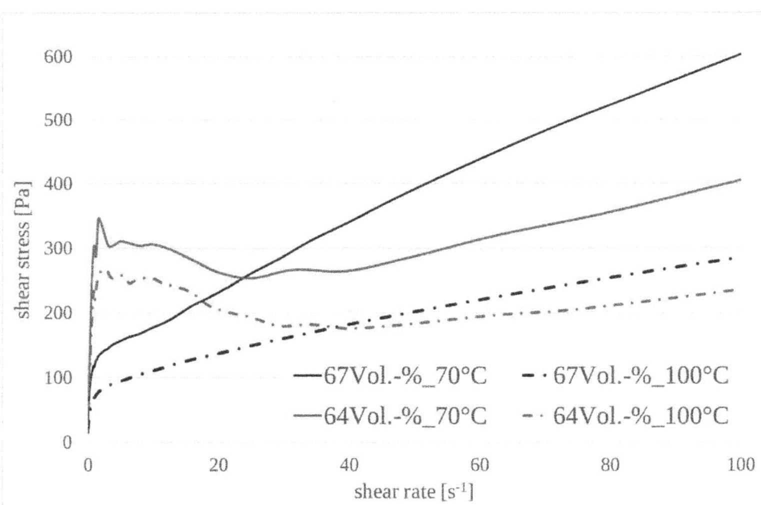

Figure 8. Shear stress vs. shear rate for two Al_2O_3-suspensions at different temperatures.

The inhomogeneous shear stress at low shear rates for the suspension with 64 vol.-% particle content, visualized inform of decreasing and alternating trends, confirm the theory of slipping. Based on the falling shear stresses the viscosity has to increase and later at higher shear rates, leading to increasing shear stresses, the viscosity is decreasing lower compared to the suspensions with 67 vol.-% particle content.

Slipping during the measurement could maybe avoid by using a measurement system with a higher surface roughness. But such a system is not available for our rheometer.

In the second test series with a suspension consisting of 64 vol.-% solid content, a filling of the micro dispensing system chamber was possible. Figure 9 shows single droplets, which were formed and deposit on a glassy substrate. Suspension reservoir and nozzle were heated up to 115°C and the climate within the laboratory was 21 °C and 36 % humidity. More than 1000 droplets could be deposited with a frequency of 200 shots/s. The droplets had a mean diameter of about 600 µm and a mean mass of about 45 µg. No defects like satellite droplets could be detected.

These tests show the possibility to form small single droplets of highly filled suspensions.

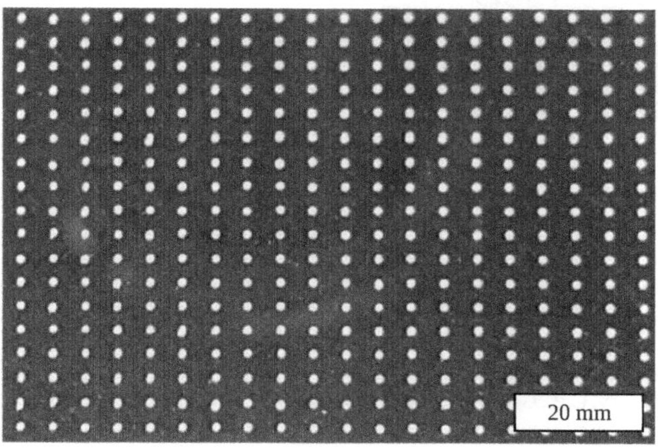

Figure 9. Single droplets after deposition with heatable micro dispensing system.

CONCLUSION

Thermoplastic 3D printing is a promising method to produce multi-material-components, e.g. steel-ceramic composites, with high geometric complexity and additionally with the option of three-dimensional functionally graded material properties.

Different suspensions have been produced for alumina, zirconia, and stainless steel. A powder content of 45 vol.-% and more can be realized for all suspensions with a maximum of 67 vol.-% for alumina. Densities of 99 % T.D. and higher were achieved for all suspensions after printing and sintering. Steel-zirconia-composits were manufactured, co-debinded and co-sintered (Figure 10) to demonstrate that T3DP can be used for AM of multi-material-components.

The adjustment of the rheological behavior of the suspensions is possible by the variation of the powder content. This is important for the deposition of small droplets. With the used dispensing equipment it was possible to deposit single droplets with a mean diameter of about 600 µm and a mean mass of about 45 µg without any dispensing failures like satellite droplets.

These promising results show that suspensions can be deposited either as small single droplets or as filaments. The combination of both dispensing technologies enables manufacturing of very thin and filigree structures and helps increasing the productivity of large volume parts of the same material significantly.

Figure 10. Cross-section of co-sintered steel-zirconia-composite made by T3DP.

REFERENCES

[1] Lee, H. C., Potapova, Y., & Lee, D. (2012). A core-shell structured, metal-ceramic composite supported Ru catalyst for methane steam reforming. *J of Power Sources, 216,* 256-260.

[2] Molin, S., Tolczyk, M., Gazda, M., & Jasinski, P. (2011). Stainless steel/yttria stabilized zirconia composite supported solid oxide fuel cell. *J. Fuel Cell Sci. Technol., 8,* 051019-1 - 051019-5.

[3] Roberts H. W., Berzins D. W., Moore B. K., & Charlton D. G. (2009). Metal-Ceramic Alloys in Dentistry: A Review. *Journal of Prosthodontics, 18, Issue 2,* 188–194.

[4] Largiller, G., Bouvard, D., Carry, C.P., Gabriel, A., Müller, J., & Staab, C. (2012). Deformation and cracking during sintering of bimaterial components processed from ceramic and metal powder mixes. Part I: Experimental investigation. *Mechanics of Materials 53,* 123-131.

[5] Guillon, O., Bordia, R.K., & Martin, C.L. (2010). Sintering of thin films/constrained sintering. In: Sintering of advanced materials (ed. Fang, Z. Z.). *Woodhead Publishing.*

[6] ASTM-Standard F2792 -12a (2012). Standard Terminology for Additive Manufacturing Technologies. *ASTM International Distributed under ASTM license by Beuth publisher.*

[7] Lakshminarayan, U., Ogrydiziak, S., & Marcus, H.L. (1990). Selective lasersintering of ceramic Materials. *Proceedings of Solid Free-Form Symposium,* 16-26.

[8] Lauder, A., Cima, M. J., Sachs, E., & Fan, T. (1992). Three dimensional printing: surface finish and microstructure of rapid prototyped components. *Materials Research Society Symposium Proceedings, 249,* 331-336.

[9] Chartier, T., & Badev, A. (2013). Rapid Prototyping of Ceramics. In: Handbook of Advanced Ceramics. *Elsevier, Oxford, UK.*

[10] Travitzky, N., Bonet, A., Dermeik, B., Fey, T., Filbert-Demut, I., Schlier, L., Schlordt, T., & Greil, P. (2014). Additive Manufacturing of ceramic-based materials. *Advanced Engineering Materials, 16,* 729-754.

[11] Pham-Gia, K., Rossner, W., Wessler, B., Schäfer, M., & Schwarz, M. (2006). Rapid Prototyping of high-density alumina ceramics using stereolithography. *cfi/Ber. DKG, 83,* 36-40.

[12] Chartier, T., Duterte, C., Delhote, N., Baillargeat, D., Verdeyme, S., Delage, C., & Chaput, C.J. (2008). Fabrication of millimeter wave components via ceramic stereo- and microstereolithography processes. *J. Am. Ceram. Soc., 91,* 2469–2474.
[13] Griffith, M.L., & Halloran, J. W. (1996). Freeform fabrication of ceramics via stereolithography. *J.*
Am. Ceram. Soc., 79, 2601-2608.
[14] Licciulli, A., Corcione, C.E., Greco A., Amicarelli, V., & Maffezzoli, A. (2005) Laser stereolithography of ZrO_2 toughened Al_2O_3. *J. Europ. Ceram. Soc., 25,* 1581-1589.
[15] Felzmann, R., Gruber, S., Mitteramskogler, G., Tesavibul, P., Boccaccini, A.R., Liska, R., & Stampfl, J. (2012). Lithography-based additive manufacturing of cellular ceramic structures. *Adv. Eng. Mater., 14,* 1052-1058.
[16] Lenk, R., Nagy, A., Richter, H.-J., & Techel, A. (2006). Material development for laser sintering of silicon carbide. *cfi/ Ber. DKG, 83,* 41-43.
[17] Regenfuss, P., Ebert, R., & Exner, H. (2007). Laser Micro Sintering - a versatile instrument for the generation of microparts. *Laser Technik Journal, 4,* 26-31.
[18] Hagedorn, Y.-C., Wilkes, J., Meiners, W., Wissenbach, K., & Poprawe. R. (2010). Net shaped high performance oxide ceramic parts by selective laser melting. *Phys. Proced., 5,* 587–594.
[19] Wu, Y., Du, J., Choy, K.-L., & Hench, L.L.: Laser densification of alumina powder beds generated using aerosol spray deposition. *J. Europ. Ceram. Soc., 27,* 4727-4735.
[20] Goodridge, R.D., Lorrison, J.C., Dalgarno, K.W., & Wood, D.J. (2004). Comparison of direct and indirect selective laser sintering of porous apatite mullite glass ceramics. *Glass Technology, 45,* 94-96.
[21] Polsakiewicz, D., & Kollenberg, W. (2012). Process and materials development for functionalized printing in three dimensions (FP-3D). *Refractories WORLDFORUM, 4,* 1-8.
[22] Sadeghian, Z., Heinrich, J. G., & Moztarzadeh, F. (2004). Direct Laser Sintering of Hydroxyapatite Implants by Layerwise Slurry Deposition (LSD). *cfi/Ber. DKG 82,* E1-E5.
[23] Cappi, B., Oezkol, E., Ebert, J., & Telle, R. (2008). Direct inkjet printing of Si_3N_4 Characterization of ink, green bodies, and microstructure. *J. Europ. Ceram. Soc., 28,* 2625-2628.
[24] Ebert, J., Özkol, E., Zeichner, A., Uibel, K., Weiss, Ö., Koops, U., Telle, R., Fischer, H. (2009). Direct Inkjet printing of dental prostheses made of zirconia. *J. Dent. Res., 88,* 673-676.
[25] Allahverdi, M., Danforth, S.C., Jafari. M., & Safari, A. (2001). Processing of advanced electroceramic components by fused deposition technique. *J. Europ. Ceram. Soc., 21,* 1485-1490.
[26] Bose, S., Darsell, J., Hosick, H., Yang, L., Sarkar, D.K., & Bandyopadhyay, A. (2002). Processing and characterization of porous alumina scaffolds. *J. Mater. Sci.: Mater. Med., 13,* 23-28.
[27] Schlordt, T., Schwanke, S., Keppner, F., Fey, T., Travitzky, N., & Greil. P. (2013). Robocasting of alumina hollow filament lattice structures. *J. Europ. Ceram. Soc., 33,* 3243–3248.
[28] Cai, K., Roman-Manso, B., Smay, J.E., Zhou, J., Osendi, M.I., Belmonte, M., & Miranzo, P. (2012). Geometrically complex silicon carbide structures fabricated by robocasting. *J. Am. Ceram. Soc., 95,* 2660-2666.
[29] Cetinel, F.A., Bauer, W., Mueller, M., Knitter. R., & Hausselt, J. (2010). Influence of dispersant, storage time and temperature on the rheological properties of zirconia-paraffin feedstocks for LPIM. *J. Europ. Ceram. Soc., 30,* 1391-1400.
[30] Gorjan, L., Dakskobler, A., & Kosmac, T. (2012). Strength evolution of injection-molded ceramic parts during wick-debinding. *J. Am. Ceram. Soc., 95,* 188-193.
[31] Scheithauer, U., Schwarzer, E., Richter, H.-J., & Moritz, T. (2014). Thermoplastic 3D Printing – An Additive Manufacturing Method for Producing Dense Ceramics, *JACT,* 1–6.

[32] Scheithauer, U., Bergner, A., Schwarzer, E., Richter, H.-J., Moritz, T. (2014). Studies on thermoplastic 3D printing of steel–zirconia composites, *J Mat Res 29 (17)*, 1931 – 1940.
[33] Scheithauer, U., Slawik, T., Schwarzer, E., Richter, H.-J., Moritz, T., & Michaelis, A. (2015). Additive Manufacturing of Metal-Ceramic-Composites by Thermoplastic 3D-Printing. *J. Ceram. Sci. Tech., 06 [02]*, 125-132.
[34] Yeo, J., Jung, Y., & Choi, S. (1998). Zirconia-stainless steel functionally graded material by tape casting. *J. Europ. Ceram. Soc., 18*, 1281-1285.
[35] Dourandish, M., Simchi, A., Shabestary, E.T., & Hartwig, T. (2008). Pressureless sintering of 3YTZP/stainless-steel composite layers. *J. Am. Ceram. Soc., 91*, 3493-3503.
[36] Dourandish, M., & Simchi, A. (2009). Study the sintering behavior of nanocrystalline 3YTZP/430L stainless-steel composite layers for co-powder injection molding. *J. Mater. Sci., 44*, 1264-1274.

MICRO-REACTORS MADE BY LITHOGRAPHY-BASED CERAMIC MANUFACTURING (LCM)

U. Scheithauer[2], E. Schwarzer[2], G. Ganzer[2], A. Körnig[2], W. Beckert[2], E. Reichelt[1], M. Jahn[2], A. Härtel[2], H.J. Richter[2], T. Moritz[2] and A. Michaelis[1,2]

[1]Institute of Materials Science, Technical University Dresden, Germany
[2]Fraunhofer Institute for Ceramic Technologies and Systems, Dresden, Germany

ABSTRACT

Heterogeneous catalysts or static mixers are widely applied in chemical industry. Regardless the extensive research efforts on the development of new catalyst supports or static mixers with low-pressure drop, the industrial application of such structures is limited to differently shaped extrudates only, especially honeycombs. However, particularly structured components made of ceramics could be advantageous in many applications, especially catalyst supports or static mixers and in small-scale decentralized plants, that have gained increasing interest in recent years, e.g. for energy generation or chemical syntheses.

Additive manufacturing techniques allow the preparation of tailor-made structures for specific reactions or applications with a high degree of freedom concerning the design. The Fraunhofer IKTS uses the lithography-based ceramic manufacturing (LCM) technology for the development of modified honeycombs with wall openings between and spoilers inside the different channels to combine enhanced mass transfer and low-pressure drop. In addition, channels with varying diameters can be realized. In this context, the application of simulation tools (multiphysics Computational Fluid Dynamic codes) and computer aided engineering is of high interest. These tools offer a cost effective "virtual" testing and characterization platform for analysis, evaluation and optimization of advantageous structures over a wide design parameter range.

The prospects of tailor-made ceramic structures as well as the potential of the LCM-technology is presented in the following study.

INTRODUCTION

Rapid prototyping (RP) or (and especially) additive manufacturing (AM), is known since more than 20 years. Nevertheless, especially for ceramics, these methods are relatively new and revolutionary procedures of tool-free shaping. The general idea of AM is the fast production of real 3D structures from solid computer models to generate prototypes without any tool[1, 2].

Today, the AM of polymers is state-of-the-art[3, 4], for metals, it finds the way into industrial fabrication[5, 6]. Many different AM techniques exist and can be classified concerning the type of build material in its pre-fabricated form (powder, liquid, solid), the material (metal, ceramic or polymer) and their physical properties or method of bonding between the layers (sintering, chemical bond). The best-known AM techniques are e.g. 3D powder printing (3DP)[7], selective laser sintering/ melting (SLS/ SLM)[8], laminated object manufacturing (LOM) or stereo lithography (STL)[9], which are in constant refinement. Actually, there are many developments in this field and many new innovative approaches dealing with multi-material manufacturing, e.g. AM multi material approaches for new applications[10, 11]. The applications range from conventional prototyping and rapid tooling to more complex, advanced applications like medical implants, and tissue engineered constructs[12-15], micro-channels[16] or various micro-systems[17-19] and a not used potential in other different applications.

Using AM as shaping technology, increases the degree of freedom concerning the design of components and enable innovative and maybe more effective structural designs, which could not be realized by conventional shaping. Until now, constructers design production-oriented and not regarding to the optimal function. However, with AM another strategy could be chosen in generating new functional or structural oriented components e.g. catalyst supports, micro reactors or static mixers. In this field of application on one side the process conditions reaction-media, temperature and pressure rule the used materials and their supports. Another important point is the ratio of surface area and pressure drop. All parameters together determine the reaction turnover respectively the quality of mixture and have to gear to each other. Hence, flow-optimized components including a high surface area and an adequate strength are necessary for this application. Conventional ceramics, formed as pellets, extruded honeycombs or foams, have shown a corresponding potential for use in this field of application.

The LCM-technology[20], an enhancement of the well-known stereo lithography, is a process for additive manufacturing of ceramics with a high surface quality and a high resolution (635 dpi). Lithoz GmbH developed this technology for the additive manufacturing of high-performance and nearly dense ceramics[21].

This paper describes the development of new, innovative flow and surface area optimized ceramic static mixers, made of alumina, beginning with an idea of a new design followed by computer-aided engineering (CAE) for flow optimization, manufacturing of structures by LCM-process followed by adjusted thermal treatment.

BASICS

Simulation and Computer-aided engineering

Computer-aided engineering (CAE) is the broad usage of computer software to aid in engineering tasks, which relate to the use of computer systems in order to establish CAD geometry and simulating the behaviour regarding fluid dynamics of designed components. In order to evaluate the quality of mixing for complex flow structures, computational fluid dynamic (CFD) models have many advantages, like better understanding of the internal mixing process, faster analysis and comparison of several geometrical and process parameters through iterative design loops.

By using commercial simulation packages (i.e. COMSOL Multiphysics and ANSYS FLUENT) or other CAE system, the mixing process for complex, three-dimensional geometries under laminar and turbulent conditions can be analyzed by import the geometry from several CAD tools (e.g. Solid Works) and later changing the CAD-data regarding to the simulations. The meshing can be performed in ANSYS ICEM-CFD environment. Accurate meshing is crucial to minimize artificial numerical diffusion, which would negatively affect the results. During the calculation, the mass and momentum balances of different Newtonian fluids (water, air ...) are solved together with an advection-diffusion equation of a passive tracer (e.g. dye). In the post-processing step, the quality of mixing can be quantified by calculating the Coefficient of Variation (CoV) described in equation 1 and the Coefficient of Distribution (CoD), shown in equation 2, of the tracer at predefined cross sections.

$$CoV = \frac{\sigma}{\bar{x}} = \frac{1}{\bar{x}}\sqrt{\frac{\sum_{i=1}^{n}(x_i - \bar{x})^2}{n}} \qquad (1)$$

The CoV describes the ratio of standard derivation to mean value with the result of an increasing mixing quality at a decreasing CoV. Thereby the equation consists of standard deviation σ, the number of values n and the mean value $\bar{x}$. An adequate advice is a value of CoV < 0.05 that means 95 % of all values are in a range of ± 10 % around the mean value.

$$CoD = \frac{CoV}{\varphi} = CoV * \frac{l}{l_{ref}}^{-1} \qquad (2)$$

The CoD, defined as ratio of CoV and φ, the ratio of length scale l and characteristic length scale l_{ref}, is an advancement of the CoV caused of dependence to the local derivation by introducing a form of gradient structure. Furthermore, the pressure drop and flow distribution can be calculated. For most applications, it is important to keep the overall pressure drop in a specified limit.

Lithography-based ceramic manufacturing – LCM

Photo polymerisation of light sensitive organic is the basis of the STL-process developed in 1986 and published in a first patent[22]. In many cases, the organic consists of monomers/ oligomers, a resin-based acrylate or a water-based acrylamide, and photoactive components. During the manufacturing of a component, shown in Figure 1, an UV-light source (laser) exposes selectively and interconnect relevant parts of an organic layer compliant to CAD-data with a defined thickness, which is controlled by a high-adjustable installation space.

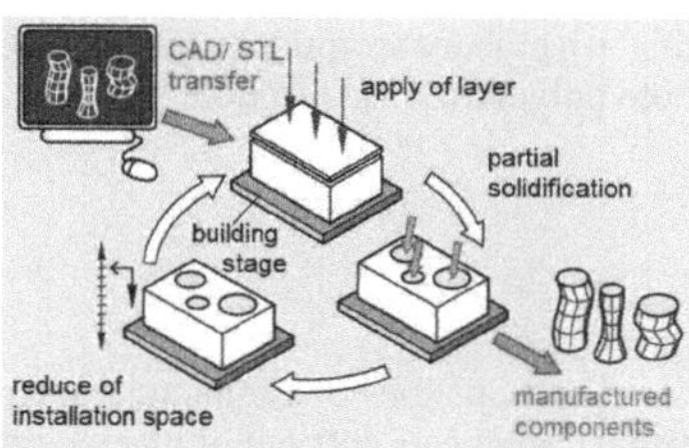

Figure 1. Schematic view of stereo-lithography from CAD/ STL to component[1].

In a further stage, the technology was developed to an area exposure method using a, for the process self-titled, digital light processing (DLP) projector. Thus, the complete part in the photoactive organic layer can be exposed simultaneously. A micro-array with two million particular controllable micro-mirrors, the core of the DLP-projector, operates in combination with an UV-light source. Figure 2 presents a schematic illustration of a micro array (A) and the DLP-process working overhead (B).

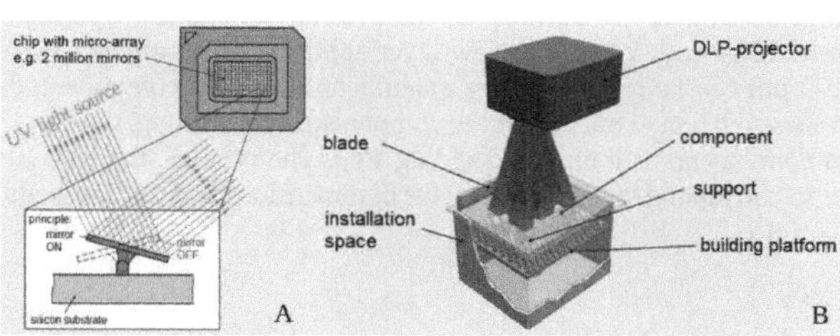

Figure 2. DLP-principle (A) and exposure in process (B)[1].

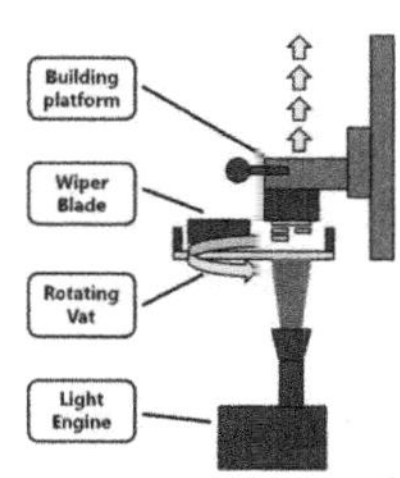

Figure 3. Scheme of the innovative overhead building process (source Lithoz).

The LCM-process derives from the DLP-process. Differences insist in a blue light source instead of UV-light combined with adjusted monomers and photoinitiators including homogeneously dispersed ceramic particles called suspension. In Figure 3 an illustration of the process of the CeraFab 7500 can be seen. The overhead working variant, an implemented new manufacturing strategy characterized by an innovative concept of lifting mechanism, is a significant advantage of this manufacturing method, compared to conventionally available systems. The exposure of the first deposited layer to light occurs by backwards located LEDs, inserted to the building platform. This cured layer will be completely adhered to the building platform and guarantee a fixation during the complete manufacturing time of the component. All subsequent layers are exposed from the bottom below the building platform by DLP irradiation throughout the glass bottom of the rotating vat. Thus, the green ceramic part, a composite of ceramic powder and organic polymer matrix, is produced backwards from the bottom to the top.

The suspension consists of at least one photoactive component in the form of photoinitiator. The photoinitiator forms free radicals, decomposed by the absorption of blue light photons at a wavelength of 465 nm, which initiate the polymerization reaction. Other ingredients could be inert additives, which absorb photons without forming free radicals or inhibitors, which stop the polymerization by reaction with free radicals[23]. In comparison to the original STL-process, the ceramic particles hinder the photo polymerization reaction, i.e. conversion, by attenuating the intensity of light source by scattering and diluting the photoactive organic[24]. The scattering is the consequence of the light reflection by ceramic particles, when the refractive index of fillers is different to the refractive index of organic matrix. For most ceramics, the refractive index is higher resulting in a higher absorption, which reduces the curing depth. Equally, smaller particles also reduce the photo polymerization reaction[25].

EXPERIMENTAL PROCEDURE

Simulation and Computer-aided engineering

The order of this study was first to assess structures and functionalities of commercial micro reactors respectively static mixers and then derive new innovative design concepts out of them which were simulated regarding to mixing quality and other flow properties to get parameters for an optimization by use of CAE. Therefore, the software tools ANSYS Fluent 14.5 and COMSOL 4.5 (CFD) were used to analyze the mixing of two fluids.

Additive Manufacturing by LCM

After accomplishing the simulations, different previously simulated designs were manufactured by LCM-technology using the CeraFab 7500. Therefore, the CAD-files, especially STL-data, were transferred to the machine. The commercially available alumina suspension Lithalox (obtainable at Lithoz) with a powder content of 35 Vol.-% has been used. The rotating vat and the viper blade apply layers in a defined thickness of 300 µm. With each new layer application a component layer thicknesses in a range of 25 to 100 µm can be realized. After manufacturing of the component, the building platform was removed followed by an extraction of the components for cleaning.

Cleaning of manufactured components is a quality-deciding step. The non-polymerized suspension has to be removed in order to correspond to the requirements of dimensions and surface quality in green and sintered state.

Thermal treatment

The components were debinded (up to 900 °C) and sintered (up to 1600 °C) in two separate steps (oven from Nabertherm) by using a defined temperature profile, given by Lithoz. After thermal processing, the different structures were characterized by computer tomography (CT) to check the quality, e.g. absence of defects and the patency of channels.

STARTING WITH AN IDEA: ABOUT SIMULATION, PRACTICAL PRODUCTION TO CHARACTERIZATION AND COMPARISON

In this study different static fluid mixers or catalyst supports, especially commercially manufactured honeycomb-structures and newly developed self-stirring mixing designs will be compared regarding their flow behavior and mixing-ability by using the CAE-Technology. CAE-Simulations allow the generation of flow-analysis before the manufacturing and testing under real conditions. In a second step, the ideas and new designs were additively manufactured by LCM process, finalized by a sintering step. Furthermore, the fired structures were characterised under real conditions in a self-constructed testing rig.

Static mixer and CAE-concepts

Instead of one single feed pipe for fluid A, which is placed in the center of the feed pipe for fluid B (Figure 4A), a honeycomb structure (Figures 4B and C) can be used where the different fluids run through the channels, in flanking channels different fluids. Structures, presented in Figures 4B and C can be manufactured by commercial shaping technologies (e.g. extrusion). Innovative designs of static mixers with adapted flow properties, could lead to a better mixing quality. The idea was to split the feed pipes to shorten the diffusion paths for the two liquids.

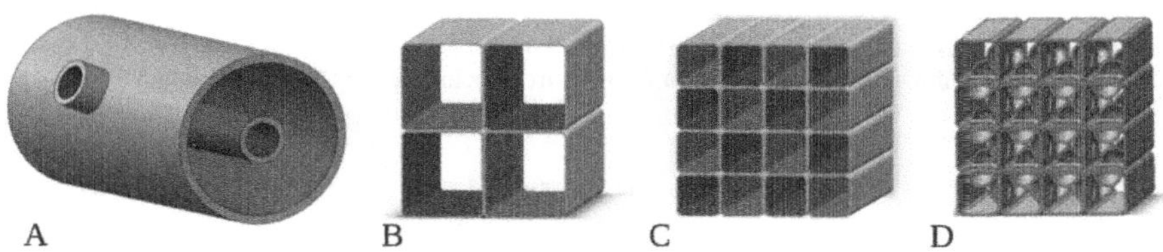

Figure 4. A: Static mixer for two fluids; B: Honeycomb with four channels as static mixer; C: honeycombs with 16 channels, D: honeycombs with 16 channels plus wall openings and spoilers.

Nevertheless the forth one (Figure 4D) is producible only by using additive manufacturing. These three designs form the basis for the simulation and comparative analysis of mixing two fluids using the software ANSYS Fluent 14.5 and COMSOL 4.5 (CFD). Figure 5 shows the scheme of simulation before mixing in structure three and the used boundary conditions.

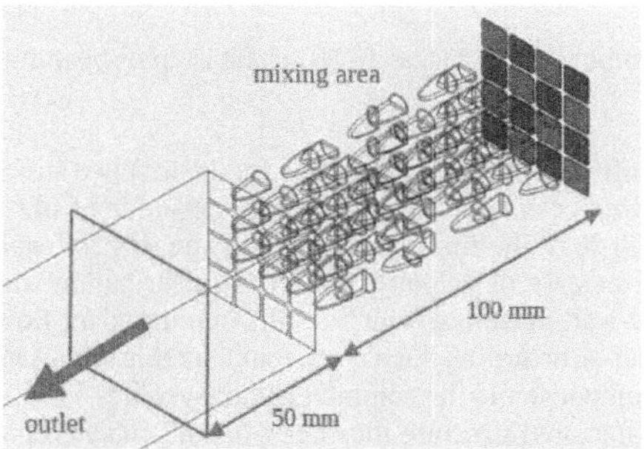

Figure 5. Scheme of initial point used for simulation of third structure, boundary conditions for both fluids: u_{in}= 1 m/s, D_{scalar} = 10^{-5} m²/s and fluid = air at RT.

Within the simulations of the fluid dynamic for the honeycomb components, significant differences regarding to flow behavior and mixing quality, after 5 and 20 cm structure length, were calculated. The difference after the two specified lengths are presented as followed. Figure 6 indicates exemplarily the simulated mixing quality and in Figure 7 the flux velocity for the first honeycomb structure is compared.

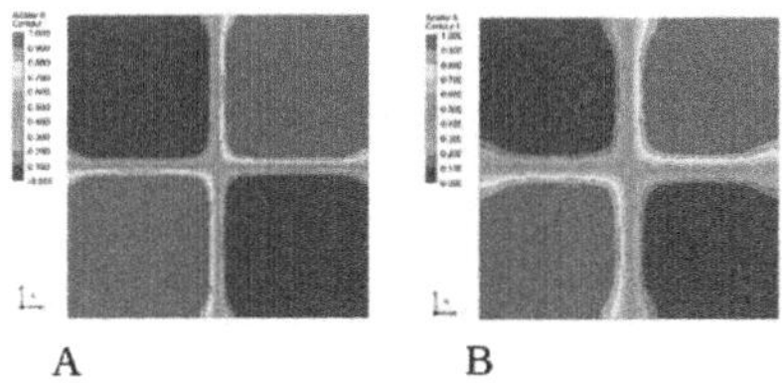

Figure 6. Mixing quality after A: 5 and B: 20 cm for structure with 4 channels.

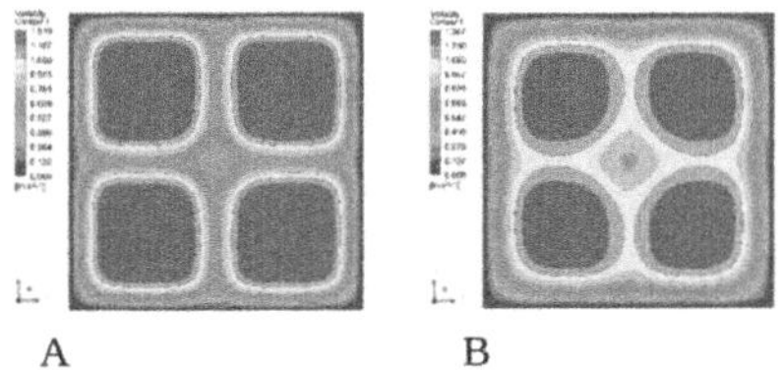

Figure 7. Flux velocity after A: 5 and B: 20 cm for a structure with 4 channels.

After 5 cm, the quality of mixture did not reach an optimum, similar to the distribution of velocity. At 20 cm, no appreciable difference to 5 cm can be seen. Considering an increase of channel number the quality of mixture and the distribution of velocity arose, but also no optimum is reached. The innovative third structure with spoilers and wall openings leads to a remarkable improvement of both factors. A comparison of all honeycomb structures, including the CoV and CoD analysis, is presented in Figure 8.

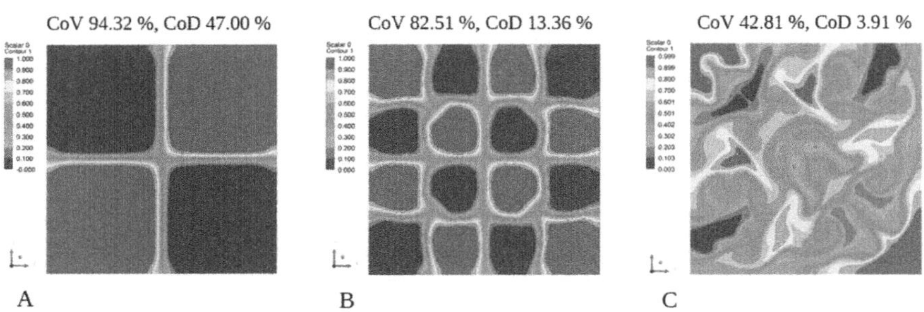

Figure 8. Simulation of mixing quality for honeycombs with 4 (A) and 16 (B) channels plus openings and spoilers (C) after a length of 5 cm.

The main reason for unequal mixture is the different number of channels for the first two structures. More channels lead to a better distribution, which can be seen in relation to the sinking CoD = 13.36 % (second structure) compared to CoD=47.00 % of the first structure. With spoilers and wall openings, realized in structure three, a significant increase of velocity and mixing distribution was calculated. The fluid needs to pass the spoilers and wall openings, which results in a turbulent flow and better mixing quality. The third innovative spoiler-structure, exclusively manufacturable with AM, demonstrates the best fluid dynamic properties in comparison to the commercial honeycombs.

A CT-image (Figure 9) of a real LCM-manufactured structure illustrates, that all channels are open, the wall openings and the spoilers were generated, too.

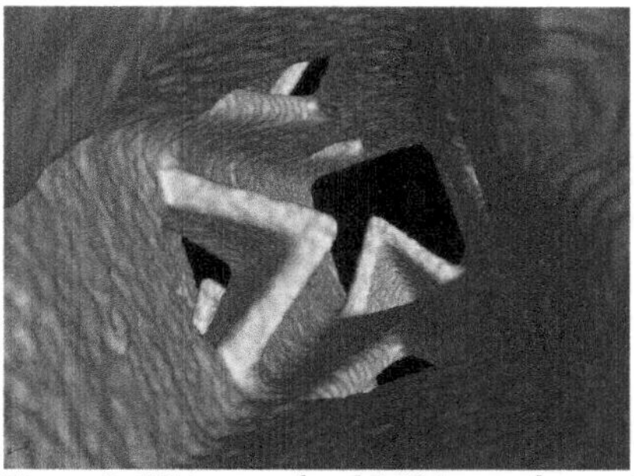

Figure 9. CT-image of real LCM-manufactured component of type three (according to Figure 4D).

A further innovative design, Figure 10, and first increment of an approach for a static mixer is a kind of a wave-like design including channels with alternating cross sections and wall openings for mixing (two) different fluids. The idea of alternating cross sections is that a part of a minimum and maximum cross section of different channels are interconnected for a mass transfer caused by a pressure difference, resulting from changing flux velocities.

Figure 10. Design of a static mixer with channels including alternating cross sections and wall openings for interconnection.

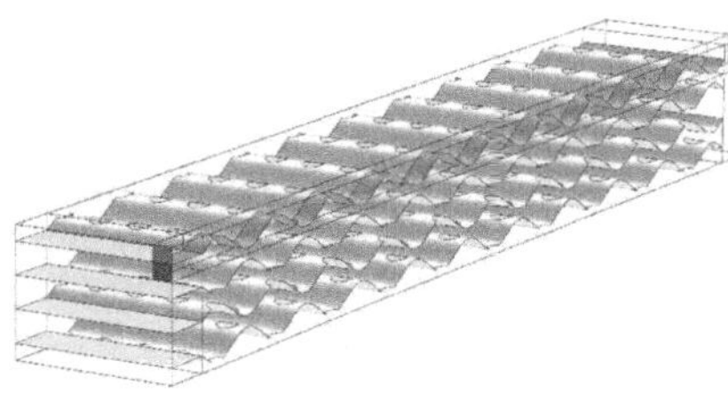

Figure 11. Geometry model and unit cell of the first increment of the described static mixer.

That means a mixing of fluids occurs due to the structural design at same flow velocities of the different fluids. Simulations concerning pressure drop, velocity and fluid concentration were carried out by converting the structure into a geometric model (Figure 11). With this boundary condition, a CFD-model (computational fluid dynamic) was developed to optimize the mixing behavior. The result of the CFD-simulations regarding to the flow and mixing behavior of these 2.5D-structures is shown in Figure 12.

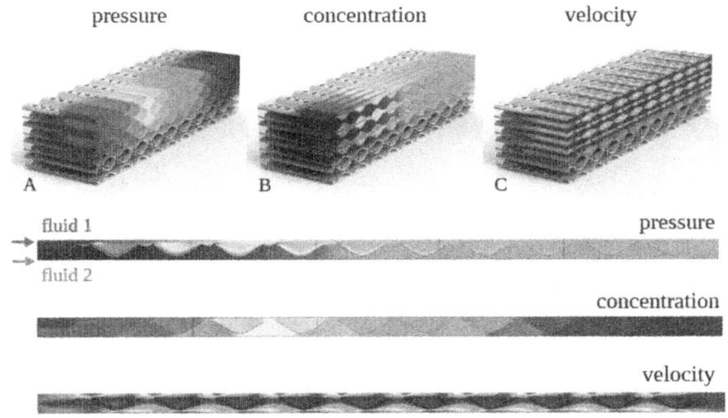

Figure 12. Results of CFD-simulation for mixing two fluids with the same properties: pressure (A), component concentration (B) and velocity (C).

The results of the simulation confirm that structures, including alternating cross sections connected by wall openings, of a static mixer have advantages compared to commercial honeycombs. Differences in flux velocity are maximal at the wall openings, in the area of minimum and maximum channel cross sections. However, there is still the challenge to realize the feeding system, which allows the homogeneous distribution of the fluids to the different channels.

A partition of one influx to a defined number (three) of channels was needed resulting in a design shown in Figure 13 (A) inform of a CFD-model and shown as construction in Figure 14.

Simulations had shown that an optimization of the design is needed in consequence of the difference of mass flow inside the channels in a range of 3.2 to 8.5 $kg/m^2 \cdot s$ due to the geometry of the inlet chamber, which is not advantageous for mixing. The construction and variation of integrated functionalities, Figure 13B, can be seen as a possible solution. Therefore, the mass flow within all channels is equal about 6.6 $kg/m^2 \cdot s$.

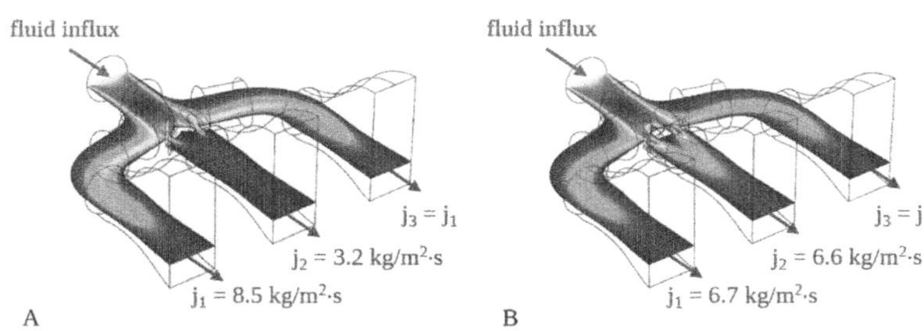

Figure 13. CFD of influx chamber (A) and optimization by integrated functionality (B)

With AM the feeding system can be manufactured with the mixing structure in one single step. That means saving of a difficult and expensive assembling step for instance by active brazing compared to conventional manufacturing. The influx velocity and volume of one fluid in different channels have to be similar to the boundary conditions of the simulation to guarantee a good match between the simulation results and the behavior of the real component. Thus, the design of the feeding system (Figure 14) can be optimized and realized with the LCM-technology.

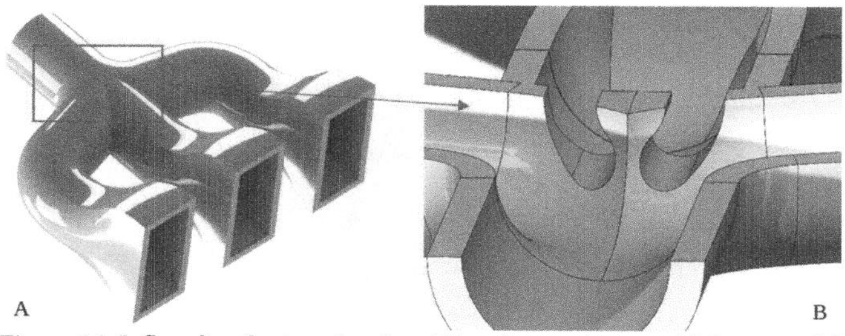

Figure 14. Influx distribution chamber (A) and components with integrated functionality (B).

The successful LCM-manufacturing and sintering, same way like first three structures of the first increment of the innovative static mixer, result shown in Figure 15, was done after simulations and optimizations compared to a component, manufactured by tape casting technology, shown in Figure 16. Both components were manufactured with an adequate quality without detecting defects.

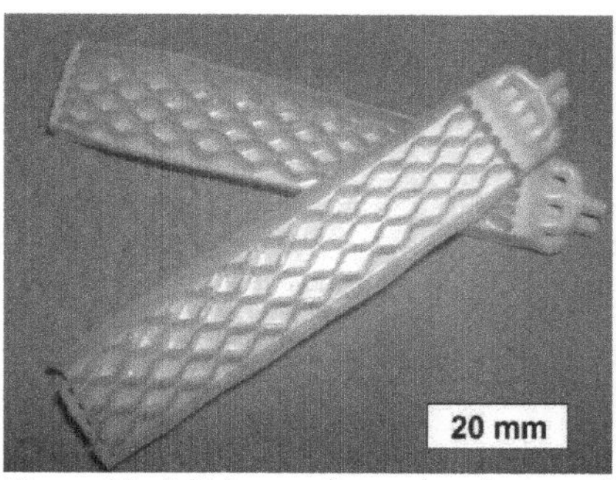

Figure 15. LCM manufactured and sintered static mixer including alternating cross sections and optimized uncompleted inlet chambers.

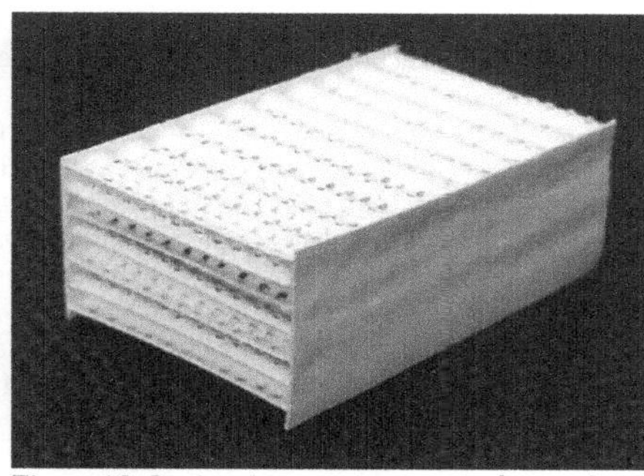

Figure 16. Static mixer structure manufactured with tape casting technology (25 x 40 x 15 mm³).

Figures 17 and 18 presents two innovative 3D-components of high performance alumina in form of different static mixers in virtual and real.

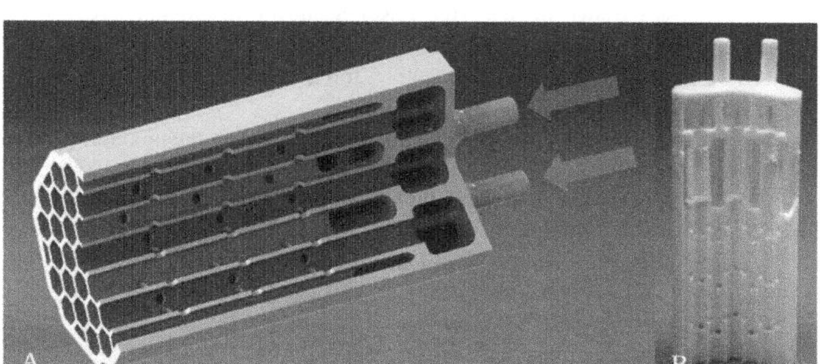

Figure 17. Sliced CAD-model (A) and additive manufactured, sintered alumina of honeycomb structure (B) with feeding system as static mixer for two different fluids.

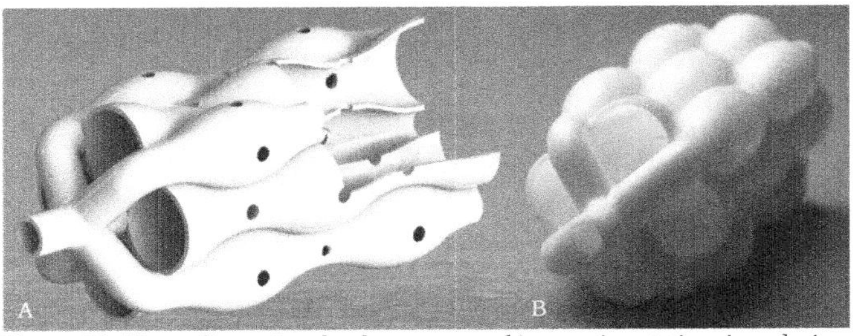

Figure 18. Three-dimensional enhancement of innovative static mixer design with changing cross sections inside the channels, presented as sliced CAD-model (A) and additive manufactured alumina part (B).

CONCLUSION

Basing on the new freedom in design, additive manufacturing by LCM, together with simulations using the method of CAE, was used to develop and to manufacture innovative designs of static mixers. The simulations regarding flow behaviour, mixing and pressure drop show remarkable differences between the geometric dimensions of the three investigated structures. A huge difference in the quality of mixing is visible and the new designed honeycomb structure with openings and spoilers, only manufacturable by AM, shows an increase of mixture quality. Basing on this result, by means of LCM, the three simulated honeycombs and several other mixers were manufactured, cleaned and finally debindered and sintered. The subsequent characterisation by CT showed the patency of the channels without remarkable defects. Developing innovative static mixture structures in combination with CAE has shown the potential of AM and used simulation tools to be a good strategy for development in this field. One advantage, which could been shown, was the simultaneous construction, optimization and manufacturing of influx chambers at the new static mixer structures for incoming fluids. Concluding, following experiments in a test rig should verify the tendencies of the fluid dynamic simulation.

REFERENCES

[1] U. Berger et al.: "Additive Fertigungsverfahren", Verlag Europa Lehrmittel (2013).
[2] J.W. Choi et al.: "Multi-material stereolithographie", J. Mat. Process. Tech., [211] 318-328 (2011).
[3] M. Hofmann, ACS Macro Lett., 3 [4] 382–386 (2014).
[4] R. D. Goodridge, C. J. Tuck and R. J. M. Hague, Prog. Mater. Sci., 57 [2] 229–267 (2012).
[5] S. Das: "Physical Aspects of Process Control in Selective Laser Sintering of Metals", Advanced Engineering Materials [5] No. 10, 2003.
[6] W. E. Frazier: "Metal Additve Manufacturing: A Review", ASM International, JMEPEG 23, 1917-1928, 2014.
[7] S. F. S. Shirazi et. al.: "A review on powder-based additive manufacturing for tissue engineering: selective laser sintering and inkjet 3D printing", Sci, Technol. Adv. Mat. 16, 20pp, 2015.
[8] J. Deckers, J. Vleugels, J.-P. Kruth: "Additive Manufacturing of Ceramics: A Review", J. Ceram. Sci. Tech., 05 [04] 245-260, 2014.
[9] U. Scheithauer, T. Slawik, E. Schwarzer, H.-J. Richter, T. Moritz, A. Michalis: "Additive Manufacturing of Metal-Ceramic-Composites by Thermoplastic 3D-Printing, J. Ceram. Sci. Tech., 06 [02] 125-132, 2015.
[10] Scheithauer, U., Slawik, T., Schwarzer, E., Richter H.-J., Michaelis, A.: Additive Manufacturing of Metal-Ceramic-Composites by Thermoplastic 3D-Printing.

[11] Scheithauer, U.; Bergner, A.; Schwarzer, E.; Richter, H.-J.; Moritz, T.: Studies on thermoplastic 3D printing of steel–zirconia composites, J Mat Res 29 (17) (2014) 1931 – 1940.
[12] B. Dhariwala et al.: "Rapid prototyping of tissue-engineering constructs, using photopolymerizable hydrogels and stereolithography", TissueEng. 10 [9–10] 1316–1322 (2004).
[13] K. Arcaute et al.: "Stereolithography of three-dimensional bioactive poly(ethylene glycol) constructs with encapsulated cells", Ann. Biomed. Eng. 34 [9] 1429–1441 (2006).
[14] K.W. Lee et al.: "Poly(propylene fumarate) bone tissue engineering scaffold fabrication using stereolithography: effects of resin formulations and laser parameters", Biomacromolecules 8 1077–1084 (2007).
[15] J.W. Choi et al.: "Cure depth control for complex 3D microstructure fabrication in dynamic mask projection microstereolithography", Rapid Prototyping J. 15 [1] 59–70 (2009a).
[16] R.B. Wicker et al.: "Embedded micro-channel fabrication using line-scan stereolithography", Assem. Autom. 25 [4] 316–329 (2005b).
[17] Y. Mizukami et al.: "A novel microchip for capillary electrophoresis with acrylic microchannel fabricated on photosensorarray", Sens. Actuators B: Chem. 81 202–209 (2002).
[18] J.A. Palmer et al.: "Mesoscale RF relay enabled by integrated rapid manufacturing", Rapid Prototyping J. 12 [3] 148–155 (2006).
[19] A. Neumeister et al.: "Generation of micro mechanical devices using stereo lithography" In: Proceedings of Solid Freeform Fabrication Symposium, The University of Texas at Austin, Austin, TX, pp. 12–24 (2006).
[20] U. K. Fischer, N. Moszner, V. Rheinberger, W. Wachter, J. Homa, W. Längle: „Lichthärtende Keramikschlicker für die stereolithographische Herstellung von hochfesten Keramiken", europäisches Patent EP 2404590A1, Veröffentlichung 11.01.2012.
[21] J. Homa Lithoz GmbH: "Rapid Prototyping of hig-performance ceramics opens new opportunities for the CIM industry", Powder Injection Moulding International, Vol. 6 No. 3, 2012.
[22] C. W. Hull: "Apparatus for production of three-dimensional objects by stereolithography", amerikanisches Patent US 4575330 (DE171069T1), Veröffentlichung 08.08.1984.
[23] Tehfe, M. A., Louradour, F. Laleveé, J., Fouassier, J.-P.: Photopolymerization Reactions: "On the Way to a Green and Sustainable Chemistry", Appl. Sci. 3, 450-514 (2013)
[24] Badev, A., Abouliatim, Y., Chartier, T., Lecamp, L., Lebaudy, P., Chaput, C., Delage, C.: Photopolymerization kinetics of a polyether acrylate in the presence of ceramic fillers used in stereolithography, J. Photoch. Photobio. A, 222, 117 – 122, 2011.
[25] Maalderink, H.: Event: High-end applications for additive manufacturing: SLA processing of highly filled resins, TNO Delft, 2013.

FUNCTIONALLY GRADED CERAMIC BASED MATERIALS USING ADDITIVE MANUFACTURING: REVIEW AND PROGRESS

[1]Li Yang, [1]Hadi Miyanaji, [2]Durga Janaki Ram, [3]Amir Zandinejad, [1]Shanshan Zhang

[1]Department of Industrial Engineering, University of Louisville, Louisville, KY, USA

[2]Department of Metallurgical and Materials Engineering, Indian Institute of Technology Madras, Madras, India

[3]Department of Oral Health and Rehabilitation, University of Louisville, Louisville, KY, USA

ABSTRACT

Functionally graded materials (FGMs) have been much sought after for use in multi-functional integrated structures. Manufacturing of these FGMs, however, remains a major technological challenge to date, especially with ceramic materials. Both ceramic-ceramic and ceramic-metal FGMs suffer from issues such as mismatch in thermal properties and improper bonding between individual layers. Additive manufacturing (AM) technologies open up exciting new possibilities in design and fabrication of FGMs with ceramic materials. This paper critically reviews the state-of-the-art in this area and provides some perspectives for future work using binder jetting 3D printing processes.

INTRODUCTION

Functionally graded materials (FGM) are characterized by gradual changes of their material compositions or structures over volumes, which usually result in gradual changes of their physical and mechanical properties [1, 2]. In recent years, significant amount of research activities have taken place in designing and understanding FGMs due to their promising potentials in many applications [3-5]. It is generally understood that FGMs bring about unprecedented design flexibility with structures by allowing for localized performance optimization and a combination of properties not achievable in regular materials [4]. For example, due to the lack of defined interface and the consequent gradual change of thermal expansion coefficients, ceramic-metal FGMs could potentially reduce the thermal-residual stresses introduced by the heating processes, and be used as a transition component for the bonding between the two dissimilar materials [1, 6]. Similarly, FGMs with graded porosities and changing material compositions could be used to achieve the intended biocompatibility and osseointegration for biomedical implants while exhibiting optimized mechanical behaviors [7, 8]. There exist two basic types of FGMs, namely the continuous FGM and the stepwise FGM as shown in Fig.1 [9, 10]. While the continuous FGMs may exhibit minimized global interfacial effects (e.g. interfacial bonding issue, mismatch of physical properties, etc.), in practice they are also significantly more difficult to produce, since the realization of continuous gradient often requires accurate molecular-level material composition control. On the other hand, although the stepwise FGMs could not eliminate global interfacial effects, they are easier to design and fabricate [2]. In addition, the interfacial effects of the stepwise FGMs could also be controlled via design optimization [11, 12].

a. Continuous FGM b. stepwise FGM

Fig.1 Basic types of FGM

FGMs are especially suitable for ceramic based materials and structures due to their potential to overcome some of the limitations in the applications. One of the most limiting issues with ceramic structures is their low damage tolerance and fracture toughness, which could be caused by internal defects as well as their interactions with the other materials [13, 14]. Many of the applications of ceramic involves high temperature environments, under which ceramics generally exhibit significantly higher structural performance over metals or polymers. However, due to their low coefficient of thermal expansion (CTE) compared to other materials and especially metals, ceramics tend to cause more significant thermal residual stresses when used jointly with the other materials, which often causes premature failure of the ceramic part of the structures [14]. By implementing FGM designs, it becomes possible to develop ceramic-metal composite structures without introducing significant internal stress. In addition, FGM designs could also improve the bonding strength between ceramics and the other types of materials, since the difference in chemical bonding often makes it difficult to create strong bonding between the two materials with a sharp interface [14].

Various manufacturing methods have been utilized to fabricate FGMs, and most of which can be classified into three categories, namely constitutive, homogenizing and segregating processes [2]. Constitutive processes build up graded structures in a stepwise manner, as is the case for powder metallurgy and coating based processes [3, 15-18]. Homogenizing processes convert a sharp interface between two materials into a gradient by mass transport, as is the case for diffusion based FGMs [19]. Segregating processes creates gradients by converting a homogeneous material into a graded material via external field (e.g. electric field or gravity), such as the infiltration and centrifugal casting based processes [20-22]. Among these methods, homogenizing and segregating processes produce continuous FGMs, but are often subject to limitations on types of gradients that can be created [2]. In comparison, powder metallurgy based constitutive processes exhibits more flexibility with FGM designs and allow for easier property designs. A body of works have been reported to utilize sintering and hot pressing technologies in the fabrication of FGMs, in which the processes usually involve the preparation of powder mixtures with varying material compositions, the stacking of different gradients, and finally the sintering or pressurized consolidation of the stacked structures [11, 12, 15, 23, 24]. This manufacturing approach possesses many advantages inherited from traditional powder metallurgy including little waste of material, flexibility with material types and generally good dimensional accuracies [25]. On the other hand, it also suffers similar limitations such as the restrictions with some geometries (e.g. undercut and internal features) and the porosity that exists in the parts. Both issues are even more critical to ceramic based FGMs, as ceramics are difficult to process with many other traditional shaping technologies. Once the FGMs are fabricated via pressing and sintering methods, it becomes almost impossible to perform secondary processes with them.

Additive manufacturing (AM), also known as 3D printing, might provide a feasible alternative in the fabrication of stepwise FGMs with maximum geometrical flexibility. AM enables digital model-driven direct fabrication of structures on a voxel-to-voxel basis, therefore allows for localized material compositional adjustment in principle [26]. Currently there exist several AM systems that are capable of multi-material fabrication and have been used for the fabrication of FGMs, however not all of them are suitable for ceramic based FGMs [2, 27, 28] When working with ceramic materials, most of these processes can be categorized into two types: the first type performs shape creation simultaneously with the property forming, while the second type only creates ceramic green part geometries, and secondary processes are needed to achieve the densification of the ceramics. Each type of approach has their advantages and limitations when it comes to the fabrication of ceramic based FGMs. With the single-step approach, the processes often face challenges with thermal residual stress control, as the required high temperature for the densification of ceramics also results in high local temperature gradient. On the other hand, with the two-step approach, it is more challenging to achieve densification for FGMs with dissimilar materials, as it is often difficult to realize localized sintering control with the post-processes. Overall the fabrication of FGMs using AM is a very novel concept and still under preliminary research and development stage, however some of the works in this areas have shown considerable potentials of the technology.

CURRENT RESEARCH

There exist eight basic types of AM processes, including vat polymerization, material extrusion, powder bed fusion, material jetting, binder jetting, directed energy deposition and direct write [29]. All of these AM processes have been studied for the fabrication of FGMs, although not all of them are suitable for ceramic based FGMs. Two design approaches have been taken to achieve the graded functionalities, namely structural gradient and material gradient. With structural gradient approach, structures with controlled porosity are often employed to achieve graded functionality [30, 31]. By designing for different porosities at different sections of the geometries, graded mechanical properties including modulus and strength of the structures can be realized. This method has also been utilized to create the graded interface when multi-material FGMs are designed [30]. When considering the structure gradient approach, the design of the porous structures become one of the most critical factors. Currently there exist extensive design tools for structures with random porosities [32, 33]. However, design theories for non-stochastic three-dimensional cellular geometries have been relatively limited [34]. On the other hand, the material gradient approach poses more challenges on the processing technology. Despite the extensive research and development works, currently AM still offers only limited capabilities for direct fabrication of FGMs. One of the most significant limiting issues is the lack of proper format to represent the FGMs. Although there already exist abundant literature works that dedicate to the modeling and characterization of performance for FGMs [24, 35-39], currently there does not exist a file format that could efficiently represent the FGMs as integrated structures for AM [40]. Currently the de-facto standard file exchange format for most AM systems is .STL format, which represents surface geometries by using 2D triangular surfaces. As the .STL format only contains surface geometry information, it does not support additional information about the material composition or part interiors. In addition, since there exist minimum material "voxel" volume limits for all the AM systems, the material compositional gradient also need to be designed as stepwise FGMs. As a result, the AM systems have to handle FGMs as assemblies pre-aligned virtually in the design software with individual components representing discrete material compositions, which makes the design process very labor intensive as the number of discrete gradients becomes larger. Although multiple studies have attempted to address this by proposing new information exchange formats, which includes the recently proposed .AMF file format, currently no AM system supports these newer formats [40, 41].

Material extrusion

Material extrusion AM processes extrude semi-liquid materials through deposition nozzles and deposit them selectively on designated areas during the printing. One of the most common material extrusion AM process is the fused deposition modeling (FDM) process, which creates the 3D geometry by melt the polymer materials supplied in filament form and extrude the semi-molten polymers through extrusion nozzles to shape the geometries in layer-wise manner as shown in Fig.2a [42, 43]. Due to the difficulty of accurate extrusion/stop control, with most material extrusion systems, continuous material tracks are formed as a result of the coordinated actions of continuous material extrusion and print head movement as shown in Fig.2b. As a consequence, the fabricated parts often exhibit significant anisotropic properties, and the material deposition pattern becomes critical in determining the performance of the parts [44].

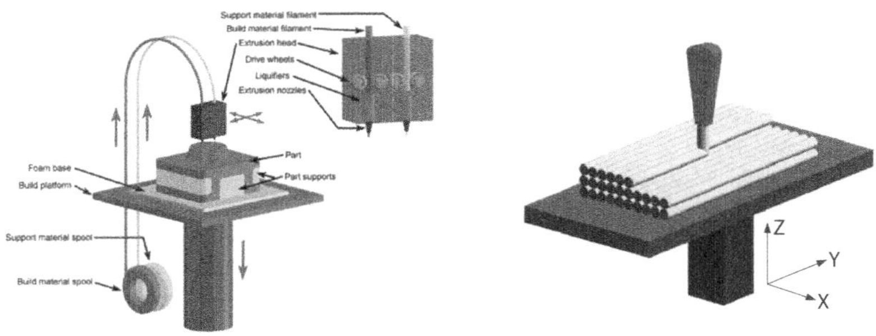

(a) Dual-material FDM [43] (b) Deposition pattern

Fig.2 Fused modeling deposition (FDM) process

Many material extrusion systems can utilize multiple materials during the printing. In principle multiple materials can be realized via implementing multiple print nozzles, although in practice many issues remain challenging such as print nozzle utilization issue and interference issue [45]. Due to the relatively easy hardware setup, multi-material printing has been widely developed for material extrusion processes. In many dual-material extrusion systems, one of the materials is often used as support structure so that the distinct difference of properties between the two materials could facilitate the easy removal of the support materials. The fabrication of FGMs with the material extrusion systems is realized through pre-mixing of multiple materials before they are co-extruded onto the substrate [46, 47]. In the case shown in Fig.3a, three materials were fed from three different extruders under separate pressure control, and are subsequently mixed in the static mixer. The extrusion pressure of each extruder need to be controlled carefully in order to ensure the accurate mixing ratio between components, and due to the fixed volume of the static mixer there exist a transition delay between the time change of compositional mixing ratio takes place and the time the mixed materials are extruded [47]. Also, due to the difficulty of mixing high viscosity materials, slurry based materials have been used for the fabrication of FGMs, which potentially limits the strength of the parts.

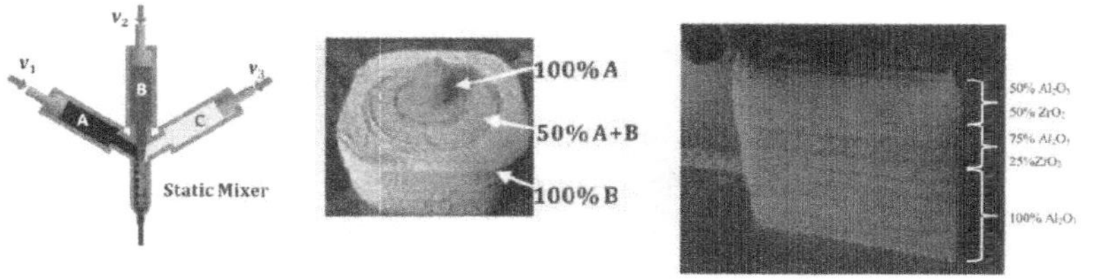

(a) Multi-material mixer   (b) $CaCO_3$ part with two colors   (c) Alumina-zirconia graded part

Fig.3 FGMs produced by material extrusion AM [47]

Using slurry based materials, material extrusion processes can be readily used to produce ceramic parts [48, 49]. On the other hand, limited studies have been carried out on the fabrication of ceramic based FGMs using this type of processes [46, 47]. As illustrated by Fig.3b and Fig.3c, FGMs various material combinations including zirconia/alumina, alumina/stainless steel and tungsten carbide have been demonstrated. One challenge of FGMs fabrication with material extrusion is that the material gradient control is limited by the material deposition pattern. Continuous material gradient can be realized via careful control of the material mixing ratio, however material gradient between the neighboring deposited tracks is discrete. In addition, with the static mixing mechanism it is also difficult to realize high responsive rate for material composition change. With the ceramic based FGMs, slurry based material extrusion becomes the only feasible deposition mode, as the melting temperature of ceramics is too high for the thermal based extrusion. As a result, the resulting solid structure are often porous after the densification because of the solid content limitation of the slurries [50].

Powder bed fusion

Powder bed fusion processes fabricate parts by selectively melting or sintering powder layer by layer in a powder bed. As shown in Fig.4, the material feedstock in the powder bed fusion processes is always supplied in layer-wise manner. After a new layer is added, the energy beam (laser or electron beam) selectively heats up the powder bed to sinter or melt it. As both sintering and melting requires significantly amount of energy input, powder bed fusion processes usually take place at elevated temperatures. Therefore, in general powder bed fusion processes have been less commonly utilized to fabricate ceramic parts. Two process approaches are usually taken when fabricating ceramic parts. The first approach pre-treats the ceramic powder by coating it with a polymer binder material [51, 52]. During the fabrication process, the polymer bincer material will be melted under the energy beam to enable liquid phase sintering of the ceramic powder. The second approach is the direct melting of ceramic powder, which usually requires high temperature preheating in order to avoid the thermal cracks that would otherwise be introduced by the drastic temperature change upon melting-solidification [53].

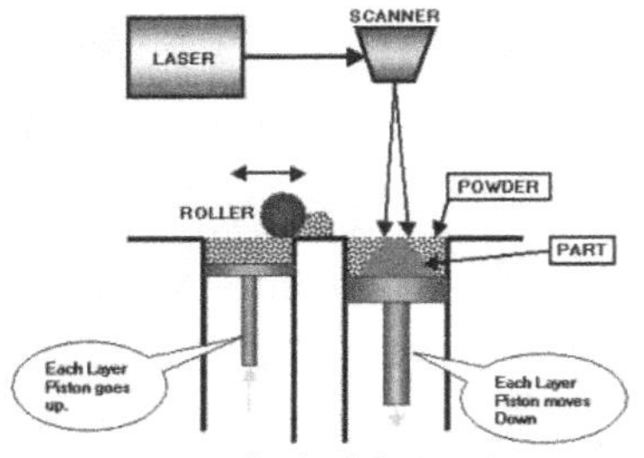

Fig.4 Laser powder bed fusion AM process

Due to the existence of powder bed, the powder bed fusion processes in general do not favor multi-material fabrication. Once multiple material powder mixes in the powder bed, it becomes very difficult to separate them. Very little works have been reported in the use of powder bed fusion processes for the fabrication of FGMs. WC/Co FGMs was fabricated via a selective laser sintering (SLS) system, however the success was limited as shown in Fig.5 [54]. Beside the difficult in thermal gradient control issue, the layer-wise material supply mechanism also makes it difficult to realize material gradient control within one layer. In fact, in order to realize gradient control, manual powder mixing was performed before the spread of each layer, which was labor intensive [54]. In general, compared to other AM processes, powder bed fusion AM processes pose more challenges in the fabrication of ceramic based FGMs.

(a) gradient parallel to build direction (b) gradient normal to build direction

Fig.5 FGMs fabricated via selective laser sintering [54]

Material jetting

Similar to the operation of a traditional inkjet printer, the material jetting processes deposit materials droplet by droplet through a printhead in order to fabricate the 3D geometries. As shown in Fig.6, multiple nozzles can be used in parallel during the printing, and multiple materials can be used simultaneously similar to the use of multi-color ink in the inkjet printers. Currently material jetting processes uses photocurable thermoset resins as the fabrication materials, therefore after the printing of each layer the materials will be cured using UV light. It has been shown that the interface between the multiple materials exhibit satisfactory tensile strength and relatively low shear strength [55]. However in general knowledge about multi-material fabrication using material jetting is still limited.

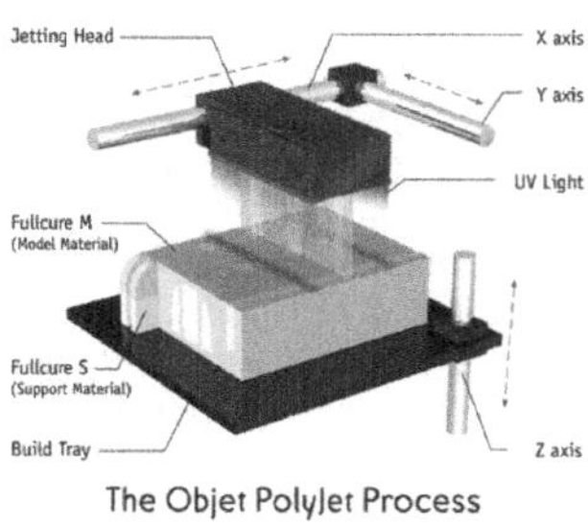

(a) Process system (b) Multi-material samples

Fig.6 Polyjet material jetting system

Although the commercially available material jetting systems are readily available for fabrication of FGMs, there has not been report on the use of this process for ceramic based FGMs. One of the limitations is the material options. In order for the process to be feasible, the solid content of ceramic in a photopolymer resin is very limited, which is due to the limitation of the overall viscosity as well as the light scattering effects resulted from the addition of ceramic solid particles [56]. Another difficulty comes from the part design. As mentioned previously, in order to design for a FGM with discrete gradient with AM, multiple parts need to be designed and virtually assembled in order to enable the direct fabrication. However, due to its ability to realize localized material composition control in all directions, it can be expected that material jetting process still possess considerable potentials in the fabrication of FGMs.

Binder jetting

Initially developed in the late 1980s, the binder jetting process works by selectively glue powder together using an organic or inorganic binder in a layer-by-layer manner as shown in Fig.7 [41, 57]. Binder jetting process realizes the fabrication of functional components by using the two-step process that creates the geometry first and then densify it to achieve part strength. In the printing process, after one layer of material powder is spread in the powder bed, a printhead with multiple nozzles enables the selective deposition of binder liquid droplets with relatively high throughput and a print resolution of 0.0635mm [58]. After the binder printing there usually exists an additional drying step before the process proceeds with the new layer. The drying can be realized via heating lamp and serves to cure the binder to achieve minimum green part strength for handling. After the entire green part is printed, it is then retrieved from the powder bed and subjected to either an infiltration or sintering to achieve higher part density and strength.

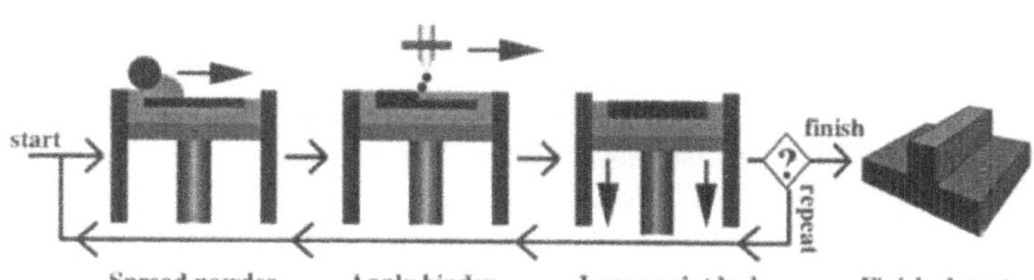

Fig.7 Binder jetting AM process [41]

Due to the layer-wise material supply, binder jetting AM has been mostly utilized to fabricate stepwise FGMs with 1-dimensional material gradients [41, 59, 60]. SEveral ceramic based FGMs have been fabricated using the binder jetting processes including SiC/Si, Al_2O_3/ZrO_2, and it was suggested that the main challenge originates from the need to perform post-sintering to the as-fabricated green parts [59, 60]. Due to the thermal expansion and shrinkage introduced during the sintering process, ceramic based FGMs fabricated via binder jetting processes will face the same challenge as the traditionally made FGMs in the control of thermal residual stress. Also, since the entire FGM part is subject to the same thermal cycle, it is also very difficult to realize the densification of ceramic-metal FGMs due to the significant difference between their sintering temperatures.

Due to the existence of powder bed, binder jetting processes are subject to similar limitations as the powder bed fusion processes in material utilization and waste. In a recent study with porcelain/Al_2O_3 FGMs, the material powder was manually changed in order to change the composition, and the unused powder in the powder bed was wasted after the fabrication [61]. On the other hand, the interfacial bonding between different compositions was shown to be satisfactory, which resulted in defect-free interface after the sintering as shown in Fig.8 [61].

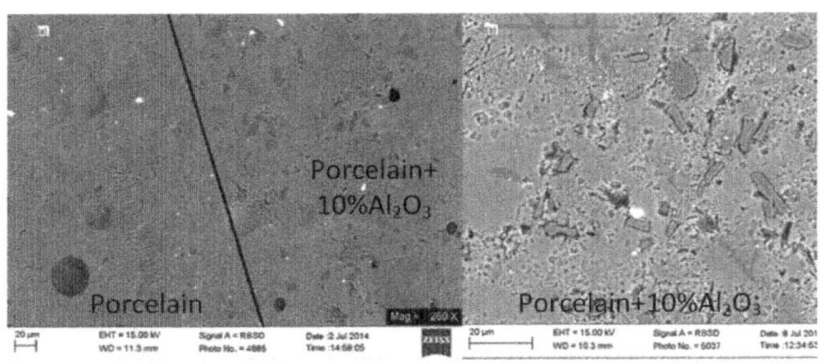

Fig.8 interface of Porcelain/Al2O3 FGM

Direct write

Direct write processes represent a number of different processes that performs high resolution direct fabrication of geometries using a drop-on-demand approach. The materials used in direct write are usually in liquid form, and the resolution of the material deposition can achieve the magnitude of pL per droplet through careful rheological control [62]. With its high resolution, direct write technology is most commonly studied for small-volume and high precision structures such as electronics, tissue scaffolds, micro-sensors and micro-actuators [62, 63]. When used for ceramic printing, direct write processes usually requires that liquids with stable ceramic particle suspension to be made. Similar to the issues encountered with material jetting, there exist a solid content limit for the direct write of certain solutions, which brings about potential issues with part strength and porosity [64]. However, as direct write can handle a wide range of liquid materials, it possesses more flexibility with material selections compared to material jetting. In addition, the drop-on-demand approach also allows for accurate voxel-to-voxel material control. Works have been reported to use direct write for the fabrication of ZrO_2/Al_2O_3 FGMs, which exhibit some porosity after the sintering [65]. Another interesting characteristics about direct write process is that it is often combined with the other AM processes in order to achieve a combined functionality [66, 67]. In these cases, the direct write acts as a local material composition control while the other processes ensures adequate fabrication speed with the bulk matrix materials. This opens up potential opportunities in the development of new process platforms that could efficiently realize high speed fabrication of FGMs.

CONCLUSION

Despite extensive research and development of the technologies, currently there does not exist an AM process that could satisfactorily fabricate ceramic based FGMs. Many challenges encountered by the traditional manufacturing of ceramic parts are also encountered by the AM processes, such as the thermal residual stress and the part porosity when slurry-based material feedstock is used. Challenge exists for both hardware and software. Some of the AM processes readily support voxel-to-voxel material compositional control, while some are restricted to layer-wise gradient control or vector-line gradient control. There is also difficulty in incorporating volume-based material information into the AM information exchange format, which to date is still heavily focused on surface geometry representation.

REFERENCES

[1] J. J. Sobczak, L. Drenchev. Metallic functionally graded materials: A specific class of advanced composites. Journal of Materials Science and Technology. 29(2013), 4: 297-316.

[2] B. Kieback, A. Newbrand, H. Riedel. Processing techniques for functionally graded materials. Materials Science and Engineering A. 362(2003): 81-105.

[3] U. Schulz, M. Peters, Fr.-W. Bach, G. Tegeder. Graded coatings for thermal, wear and corrosion barriers. Materials Science and Engineering A. 362(2003): 61-80.

[4] G. Udupa, S. Shrikantha Rao, K. V. Gangadharan. Functionally graded composite materials: An overview. Procedia Materials Science. 5(2014): 1291-1299.

[5] E. Muller, C. Drasar, J. Shilz, W. A. Kaysser. Functionally graded materials for sensor and energy applications. Materials Science and Engineering A. 362(2003): 17-39.

[6] M. Koizumi, M. Niino. Overview of FGM research in Japan. MRS Bulletin. 20(1995), 1: 19-21.

[7] W. Pompe, H. Worch, M. Epple, W. Friess, M. Gelinsky, P. Greil, U. Hempel, D. Scharnweber, K. Schulte. Functionally graded materials for biomedical applications. Materials Science and Engineering A. 362(2003): 40-60.

[8] M. Mehrali, F. S. Shirazi, M. Mehrali, H. S. C. Metselaar, N. A. B. Kadri, N A. A. Osman. Dental implants from functionally graded materials. Journal of Biomedical Materials Research A. 101A(2013), 10: 3046-3057.

[9] Y. Miyamoto, W. Kaysser, B. Rabin, A. Kawasaki, R. Ford. Functionally graded materials: design, processing and applications. Kluwer Academic Publishers, 1999.

[10] P.-H. Lee. Fabrication, Characterization and Modeling of Functionally Graded Materials. PhD Dissertation, Columbia University, New York, NY, 2013.

[11] J. C. Lee, J. H. Park, S. H. Ryu, H. J. Hong, D. H. Riu, S. H. Ahn, C. S. Lee. Reduction of functionally graded material layers for Si3N4-Al2O3 system using three-dimensional finite element modeling. Materials Transactions. 49)2008), 4: 829-834.

[12] Y. M. Shabana, H. A. Bruck, M. L. Pines, J. G. Kruft. Modeling the evolution of stress due to differential shrinkage in powder-processed functionally graded metal-ceramic composites

during pressureless sintering. International Journal of Solids and Structures. 43(2006): 7852-7868.

[13] F. Bouville, E. Maire, S. Meille, B. Van de Mooretele, A. J. Stevenson, S. Deville. Strong, tough and stiff bioinspired ceramics from brittle constituents. Nature Materials. 13(2014): 508-514.

[14] J. A. Fernie, R. A. L. Drew, K. M. Knowles. Joining of engineering ceramics. International Materials Reviews. 54(2009): 283-331.

[15] A. H. Wu, W. B. Cao, C. C. Ge, J. F. Li, A. Kawasaki. Fabrication and characteristics of plasma facing SiC/C functionally graded composite material. Materials Chemistry and Physics. 91(2005): 545-550.

[16] J. F. Groves, H. N. G. Wadley. Functionally graded materials synthesis via low vacuum directed vapor deposition. Composites Part B. 28B(1997): 57-69.

[17] W. Y. Lee, Y. W. Bae, K. L. More. Synthesis of functionally graded metal-ceramic microstructures by chemical vapor deposition. Journal of Materials Research. 10(1995), 12: 3000-3002.

[18] W. Y. Lee, D. P. Stinton, C. C. Berndt, F. Erdogan, Y.-D. Lee, Z. Mutasim. Concept of functionally graded materials for advanced thermal barrier coating applications. Journal of American Ceramic Society. 79(1996), 12: 3003-3012.

[19] W. Lengauer, K. Dreyer. Functionally graded hardmetals. Journal of Alloys and Compounds. 338(2002): 194-212.

[20] H. Hassanin, K. Jiang. Infiltration-processed, functionally graded materials for microceramic components. Micro Electro Mechanical Systems (MEMS), 2010 IEEE 23rd International Conference on, 368-371.

[21] S. Pratapa, I. M. Low, B. H. O'Conner. Infiltration-processed, functionally graded aluminium titanate/zirconia-alumina composite. Journal of Materials Science. 33(1998): 3037-3045.

[22] Y. Watanabe, N. Yamanaka, Y. Fukui. Control of composition gradient in a metal-ceramic functionally graded material manufactured by the centrifugal method. Composite Part A. 29A(1998): 595-601.

[23] S. N. S. Jamaludin, F. Mustapha, D. M. Nuruzzaman, S. N. Basri. A review on the fabrication techniques of functionally graded ceramic-metallic materials in advanced composites. Scientific Research and Essays. 8(2013), 21: 828-840.

[24] Z. He. J. Ma, G. E. B. Tan. Fabrication and characteristics of alumina-iron functionally graded materials. Journal of Alloys and Compounds. 486(2009): 815-818.

[25] M. P. Groover. Fundamentals of Modern Manufacturing: Materials, Processes, and Systems. 5th edn. Wiley, USA, 2013.

[26] I. Gibson, D. Rosen, B. Stucker. Additive Manufacturing Technologies: 3D Printing, Rapid Prototyping, and Direct Digital Manufacturing. 2nd edn. Springer-Verlag, New York, NY, USA, 2015.

[27] A. Joshi, A. Patnaik, B. Gangil, S. Kumar. Laser assisted rapid manufacturing technique for the manufacturing of functionally graded materials.

[28] J. P. Moore, C. B. Williams. Fatigue characterization of 3D printed elastomer material. 23th International Solid Freeform Fabrication (SFF) Symposium, Austin, TX, USA, 2012.

[29] I. Gibson, D. Rosen, B. Stucker. Additive Manufacturing Technologies: 3D Printing, Rapid Prototyping, and Direct Digital Manufacturing, 2nd edn. Springer, New York, NY, 2015.

[30] V. R. Gervasi, R. S. Crockett. Composites with gradient properties from solid freeform fabrication. 9th International Solid Freeform Fabrication (SFF) Symposium, Austin, TX, USA, 1998.

[31] B. Starly, J. Nam, W. Lau, W. Sun. Layered composite model for design and fabrication of bone replacement. 13th International Solid Freeform Fabrication (SFF) Symposium, Austin, TX, USA, 2002.

[32] M. F. Ashby, A. G. Evans, N. A. Fleck, L. J. Gibson, J. W. Hutchinson, H. N. G. Wadley. Metal Foams: A Design Guide. Butterworth Heinemann, 2000.

[33] L. J. Gibson, M. F. Ashby. Cellular solids: Structure and properties, 2nd edn. Cambridge University Press, 1997.

[34] L. Yang, O. Harrysson, D. Cormier, H. West, H. Gong, B. Stucker. Additive manufacturing of metal cellular structures: design and fabrication. Journal of Materials. 67(2015), 3: 608-615.

[35] M. R. Hill, W.-Y. Lin. Residual stress measurement in a ceramic-metallic graded material. Journal of Engineering Materials and Technology. 124(2992), 2: 185-191.

[36] P. Gu, R. J. Asaro. Cracks in functionally graded materials. International Journal of Solids and Structures. 34(1997), 1: 1-17.

[37] A. Radman, X. Huang, Y. M. Xie. Topology optimization of functionally graded cellular materials. Journal of Materials Science. 48(2013): 1503-1510.

[38] L.-C. Guo, N. Noda. An analytical method for thermal stresses of a functionally graded material cylindrical shell under a thermal shock. Acta Mechanica. 214(2010): 71-78.

[39] B. D. Chapman. Characterization of functionally graded materials. M.S. Thesis. U.S. Air Force Institute of Technology, 2006.

[40] J. D. Hiller, H. Lipson. STL 2.0: a proposal for a universal multi-material additive manufacturing file format. 20th International Solid Freeform Fabrication (SFF) Symposium, Austin, TX, USA, 2009.

[41] T. R. Jackson, H. Liu, N. M Patrikalakis, E. M. Sachs, M. J. Cima. Modeling and designing functionally graded material components for fabrication with local composition control. Materials and Design, 20(1999): 63-75.

[42] N. S. A. Bakar, M. R. Alkahari, H. Boejang. Analysis on fused deposition modelling performance. Journal of Zhejiang University-Science A. 11(2010), 12: 972-977.

[43] http://www.custompartnet.com/wu/fused-deposition-modeling. (Accessed 06/2015)

[44] S.-H. Ahn, M. Montero, D. Odell, S. Roundy, P. K. Wright. Anisotropic material properties of fused deposition modeling ABS. Rapid Prototyping Journal. 8(2002), 4: 248-257.

[45] J. P. Wachsmuth. Multiple independent extrusion heads for fused deposition modeling. M.S. Thesis, Virginia Polytechnic Institute and State University, Blacksburg, Virginia, 2008.

[46] G. E. Hilmas, J. L. Lombardi, R. A. Hoffman. Advances in the fabrication of functionally graded materials using extrusion freeform fabrication. Functionally Graded Materials 1996, edited by I. Shiota, M. Y. Miyamoto, Elsevier, 1997.

[47] B. K. Deuser. Process planning and control of functionally graded parts using freeze-form extrusion fabrication. M.S. Thesis, Missouri University of Science and Technology. Rolla, MO, 2012.

[48] M. A. Jafari, W. Han, F. Mohammadi, A. Safari, S. C. Danforth, N. Langrana. A novel system for fused deposition of advanced multiple ceramics. Rapid Prototyping Journal. 6(2000), 3: 161-17.

[49] S. Onagoruwa, S. Bose, A. Bandyopadhyay. Fused deposition of ceramics (FDC) and composites. 12th International Solid Freeform Fabrication (SFF) Symposium, Austin, TX, USA, 2001.

[50] M. S. Mason, T. Huang, R. G. Landers, M. C. Leu, G. E. Hilmas. Freeform extrusion of high solids loading ceramic slurries, PartI: Extrusion process modeling. 17th International Solid Freeform Fabrication (SFF) Symposium, Austin, TX, USA, 2006.

[51] E. A. Griffin, S. McMillin. Selective laser sintering and fused deposition modeling processes for functional ceramic parts. 6th International Solid Freeform Fabrication (SFF) Symposium, Austin, TX, USA, 1995.

[52] J. C. Lorrison, R. D. Goodridge, K. W. Dalgarno, D. J. Wood. 13th International Solid Freeform Fabrication (SFF) Symposium, Austin, TX, USA, 2002.

[53] Y.-C. Hagedorn, J. Wilkes, W. Meiners, K. Wissenbach, R. Poprawe. Net shaped high performance oxide ceramic parts by selective laser melting. Physics Procedia. 5(2010): 587-594.

[54] L. Jepson, J. J. Beaman, D. L Bourell, K. L. Wood. SLS processing of functionally gradient materials. 8th International Solid Freeform Fabrication (SFF) Symposium, Austin, TX, USA, 1997.

[55] J. P. Moore, C. B. Williams. Fatigue characterization of 3D printed elastomer material. 23th International Solid Freeform Fabrication (SFF) Symposium, Austin, TX, USA, 2012.

[56] G. A. Brady, T.-M. Chu, J. W. Halloran. Curing behavior of ceramic resin for stereolithography. 7th International Solid Freeform Fabrication (SFF) Symposium, Austin, TX, USA, 1996.

[57] S. Uhland, R. Holman, B. DeBear, P. Saxton, M. Cima, E. Sachs, Y. Enokido, H. Tsuchiya. Three-dimensional printing, 3DP, of electronic ceramic components. 10th International Solid Freeform Fabrication (SFF) Symposium, Austin, TX, USA, 1999.

[58] M-Flex Specification. http://www.exone.com/Portals/0/Systems/M-Flex/X1_MFlex_US.pdf. ExOne. (Accessed 06/2015)

[59] J. Moon, A. C. Caballero, L. Hozer, Y.-M. Chiang, M. J. Cima. Fabrication of functionally graded reaction infiltrated SiC-Si composite by three-dimensional printing (3DP) process. Materials Science and Engineering A. 298(2001): 110-119.

[60] J. Yoo, K. Cho, W. S. Bae, M. Cima, S. Suresh. Transformation-toughened ceramic multilayers with compositional gradients. Journal of American Ceramic Society. 81(1998), 1: 21-32.

[61] H. Miyanaji, L. Yang, S. Zhang, A. Zandinejad. A preliminary study of the graded dental porcelain ceramic structures fabricated via binder jetting 3D printing. 25th International Solid Freeform Fabrication (SFF) Symposium, Austin, TX, USA, 2014.

[62] J. A. Lewis. Direct ink writing of 3D functional materials. Advanced Functional Materials. 16(2006): 2193-2204.

[63] K. Church. Fine lie direct printing for electronics and 3D prints. RAPID 2012, Minneapolis, MN, 2012.

[64] J. Ebert, E. Ozkol, A. Zeichner, K. Uibel, O. Weiss, U. Koops, R. Telle, H. Fischer. Direct inkjet printing of dental prostheses made of zirconia. Journal of Dental Research. 88(2009): 673-676.

[65] M. Mott, J. R. G. Evans. Zirconia/alumina functionally graded material made by ceramic ink jet printing. Materials Science and Engineering A. 271(1999): 344-352.

[66] F. Medina, A. Lopes, A. Inamadar, R. Hennessey, J. Palmer, B. Chavez, D. Davis, P. Gallegos, R. Wicker. Hybrid manufacturing: integrating direct and stereolithography. 16th International Solid Freeform Fabrication (SFF) Symposium, Austin, TX, USA, 2005.

[67] H. R. Thomas, N. Hopkinson, P. Erasenthiran. High speed sintering- continuing research into a new rapid manufacturing process. 16th International Solid Freeform Fabrication (SFF) Symposium, Austin, TX, USA, 2005.

Advanced Materials, Technologies, and Devices for Electro-Optical and Biomedical Applications

A NEUTRON DETECTOR BASED ON BORON-10 ENRICHED SCINTILLATING GLASSES

Dat Vu[1,*], Makena Dettmann[1,*], Victor Herrig[1,*], Luiz G. Jacobsohn[2], Matthew W. Kielty[2], James Wetzel[3], Yasar Onel[3], and Ugur Akgun

[1] Physics Department, Coe College, Cedar Rapids, IA 52402, USA
[2] Department of Materials Science and Engineering, and COMSET - Center for Optical Materials Science and Engineering Technologies, Clemson University, Clemson, SC 29634, USA
[3] Department of Physics and Astronomy, University of Iowa, Iowa City, IA 52242, USA

*) These authors contributed equally to the work

ABSTRACT

Neutron detectors have applications in fields ranging from oil drilling to homeland security. A design for a compact glass detector capable of measuring neutrons with energies from several MeV down to thermal energies is described. The detector described uses a combination of polyethylene layers and layers of Cerium-doped scintillating glass enriched with ^{10}B in order to slow the fast neutrons and capture them at thermal energy levels. The ^{10}B-enriched glass samples were produced and their light absorption, emission and scintillation yield properties were measured. GEANT4 software was then used to simulate a prototype detector's performance. The geometrical optimization and neutron detection capabilities were also studied.

1. INTRODUCTION

Since the neutron is electrically neutral it cannot be detected directly. Recoiled protons in the medium and secondary particle emission after a neutron capture provide a mechanism for indirect detection of neutrons with energies on the order of several MeV. Neutron will scatter repeatedly from protons, which carry away one-half of the initial energy on average. In a hydrogenous medium with a size larger than the neutron mean-free-path, these interactions deliver a series of scintillation pulses which are reduced after each scatter by one-half[1]. The recoiled proton is a localized point of ionization (the range of a 3 MeV proton in a plastic scintillator is ~10 μm), and is therefore essentially a point source of luminescence. Thus, the physical signal is a sequence of point sources of light, each dimmer by a factor of two. On average, these points will align with the original neutron trajectory[2].

Excellent signatures for thermal neutrons are generated by several neutron capture reactions. Since Boron oxide is a well-known glass-former and the de-excitation signal is substantially large, this study concentrates on neutron capture in ^{10}B via the following reactions:

^{10}B + n --> Li + α (1.78 MeV) - 6% probability
^{10}B + n --> ^{7}Li + α (1.47 MeV) + gamma (0:48 MeV) - 94% probability

The polyethylene layers detailed in the detector's design increase the Hydrogen content and provide an ideal medium to slow down the fast neutrons. The cross section for ^{10}B increases drastically and reaches 4000 barn at thermal energy levels. The signals from the neutron capture—an alpha particle or a 2.3 MeV gamma ray that Compton scatters to yield an ionizing electron—will generate the scintillation photons to be detected by the photomultiplier tubes (PMTs) mounted on the side of the sample. In the sections below we describe the glass-forming conditions of Boron-10-enriched Cerium-doped scintillating glasses, the tests on their light production capabilities, and the simulation studies on a possible prototype with these glasses.

2. ^{10}B GLASS PREPARATION

^{10}B-based glass scintillators were first studied with a focus on glasses containing Boron, Sodium, Aluminum, and Cerium as the scintillator[3]. In a separate study, it was also shown that glasses with a high Boron content increase in efficiency with higher Boron concentration in glass[4]. This shows that Cerium-doped Borosilicate glasses, even with natural Boron, can be effectively used for neutron detection[4].

Since these initial results many following studies in the area have focused on ^{6}Li as the neutron-capture mechanism due to the significant α-particle peak in ^{10}B events. However, in the thermal energy region, the neutron-capturing cross-section of ^{10}B is over four times larger than that of ^{6}Li, which makes ^{10}B glasses a crucial candidate. ZnS:Ag/$^{10}B_2O_3$ glass scintillator was proved to provide a better neutron detection efficiency than ^{6}Li glasses, 61% vs. 40%, in thermal energy region[5]. ^{10}B glasses are also verified to have a significantly shorter decay time than similar ^{6}Li glasses[6]. However, increasing the ^{10}B concentration within the glass lowered the scintillating light production[7].

In the study described here, a soda-lime borosilicate glass was chosen due to its well-studied features, such as low refractive index and high transparency that make it a good candidate for adding a cerium dopant. The glass-forming conditions of Boron oxide are well studied[8,9,10]. This investigation focused on maximizing the Boron content, while yielding the most light out of Cerium scintillation. Initially, scintillating glasses with %70 B_2O_3 content were successfully made (Table 1). These glasses were found to be highly hygroscopic, drastically decreasing their optical clarity.

In order to make the samples more stable, the Boron content was reduced and glass samples were made with SiO_2 ranging from (molar %) 42.7% - 72.7%, B_2O_3 ranging from 0% - 30%, 14% Na_2O, 9% CaO, 4% MgO, .15% Al_2O_3, .1% Fe_2O_3, .03% K_2O, .02% TiO_2 and Ce dopant ranging from 0 - 5%. The composition of the final glass sample was 62.7SiO_2-14Na_20-10B_2O_3-9CaO-4MgO-.15Al_2O_3-.1Fe_2O_3-.03K_2O-.02TiO_2 + .01Ce (molar %), using the precursors $CeCl_3$, Na_2CO_3, 2H_3BO_3, SiO_2, Al_2O_3, CaO, MgO, K_2CO_3, TiO_2 and Fe_2O_3, all of at least 99% purity. Samples with both ^{10}B and natural Boron were prepared for tests. The components were individually weighed in a glass beaker, zeroing the scale after each component was added. Chemicals were added in order of least to most contribution to the total mass. The Cerium, which came in small pellets, needed to be thoroughly crushed with a mortar and pestle before adding and weighing it. After all the components were added. The powder was mixed for ten minutes with a stirrer, then three small platinum crucibles were filled with a portion of the mix. The crucibles were heated in the same furnace for 15 minutes at 1550 °C. After the first three batches were heated and cooled sufficiently, more powder was added to each crucible and the crucibles were heated again in the same manner until no powder remained. Immediately after the last heating all the crucibles were simultaneously poured into a 5 cm x 5 cm x 1 cm steel mold pre-heated to 300 °C. The steel mold was then annealed on a hot plate at 300° C for 30 minutes. The resulting plates of glass were tested for structural integrity. The protruding parts were sanded away and polished.

Sample	cut-off (nm)	Li_2O	SiO_2	Al_2O_3	B_2O_3	Ce^{3+}	Na_2O	K_2O	Rb_2O	Cs_2O
1	378	20	8	1	70	1	0	0	0	0
2	367	0	8	1	70	1	20	0	0	0
3	383	0	8	1	70	1	0	20	0	0
4	373	0	8	1	70	1	0	0	20	0
5	391	0	8	1	70	1	0	0	0	20
6	435	12	12	1	70	5	0	0	0	0
7	338	14.25	14.5	1	70	0.25	0	0	0	0
8	323	14	14.5	1	70	0.5	0	0	0	0
9	379	13.75	14.5	1	70	0.75	0	0	0	0
10	326	14	14	1	70	1	0	0	0	0
11	314	20.8	8	1	70	0.2	0	0	0	0
12	343	20.8	8	1	*70	0.2	0	0	0	0
13	404	13	14	1	70	2	0	0	0	0

* ^{10}B-Enriched

Table 1: The 70% boric acid samples that were produced for this study.

3. OPTICAL AND BEAM TESTS OF THE GLASS

Ideally, the emission peak of the glass would be in a region of high transmission. Cerium is known to have host-dependent scintillation properties due to its valance electron energy variation. The final sample was found to have light emission wavelengths from 325 nm to 500 nm with peak at 392 nm. Tests using the UV-VIS-NIR spectrometer revealed the light transmission of the sample to be ~50% at this peak emission wavelength (Figure 1). Although it is not the optimal transmission, this much light emission should be enough for particle detectors using high quantum efficiency photomultiplier tubes.

The samples were then taken to beam tests at CERN and the Fermilab Test Beam Facility. At these facilities the light production capabilities of the samples were tested with 150 GeV electron and muon beams (CERN), and a 120 GeV proton beam (Fermilab) (Figures 2 and 3). The scope view of the signal and the beam tests showed a decay constant of 20 ns for our glass sample. During the passage of a minimum ionizing particle, the light yields of the scintillating glasses were compared to that of a known plastic scintillator (Saint Gobain BC-412). The calculated light production capability of our glass sample was ~10,000 photons/MeV (see Figure 4).

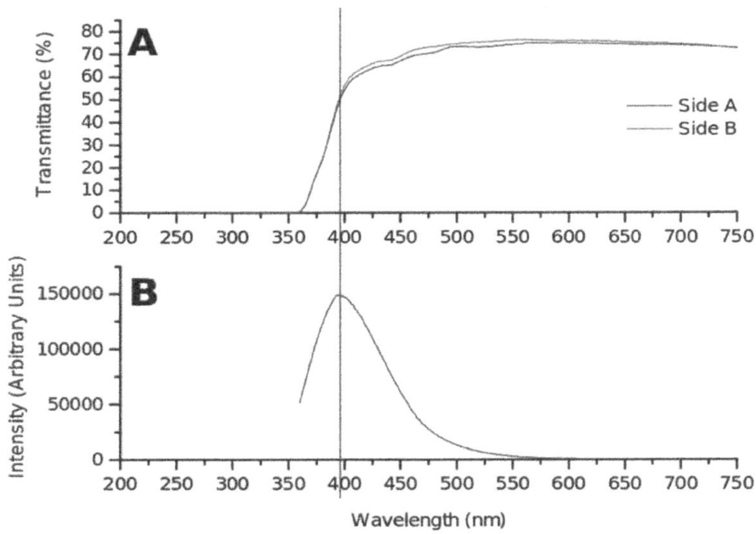

Figure 1. A. The transmission spectra of the scintillating glass samples measured with UV/Vis/NIR B. Photoluminescence emission spectra obtained from excitation at 340 nm. Side A and B corresponds to two sides of the glass slab.

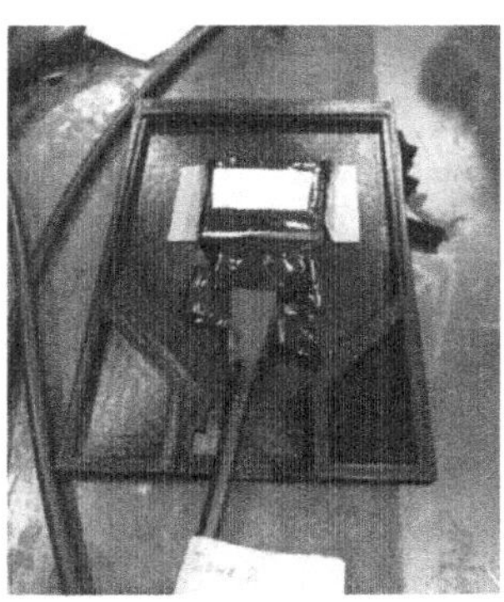

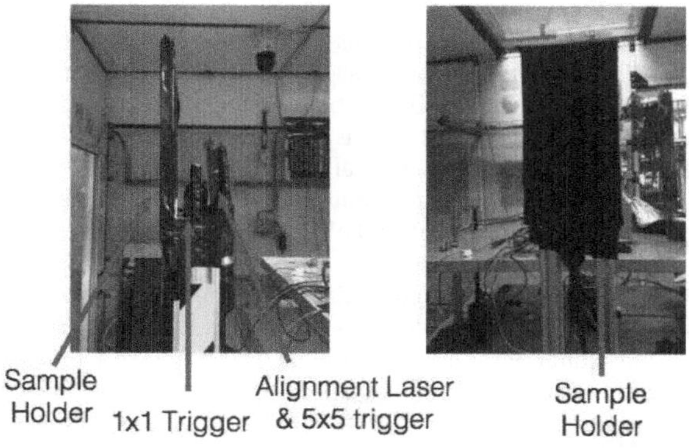

Figure 2. 3D printed, light-tight sample holders for beam tests (LEFT). The Fermilab test beam setup (RIGHT, MIDDLE).

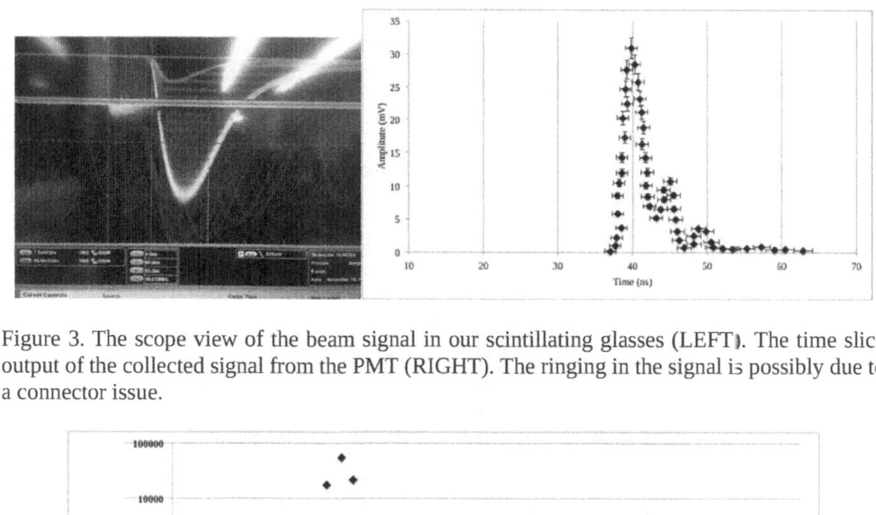

Figure 3. The scope view of the beam signal in our scintillating glasses (LEFT). The time slice output of the collected signal from the PMT (RIGHT). The ringing in the signal is possibly due to a connector issue.

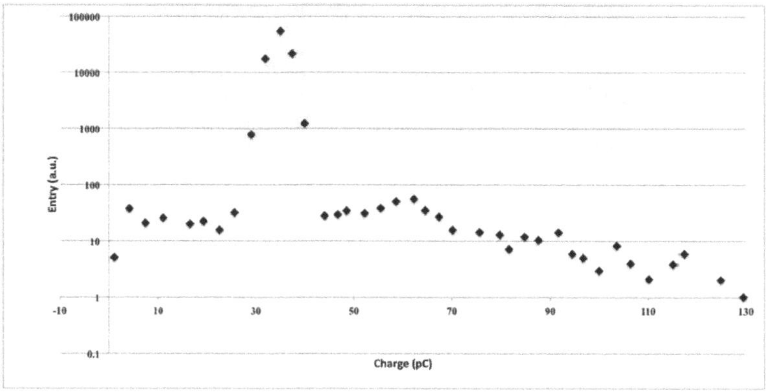

Figure 4: The ADC output of the collected signal from PMT.

4. DETECTOR PERFORMANCE STUDIES

The geometry optimization and detection efficiency of a possible prototype were simulated using Geant4[11,12], a software toolkit for the simulation of the passage of particles through matter. The detector prototype design includes 15 layers of scintillating borosilicate glasses coupled to polyethylene slabs (which act as thermalizers). In order to be as accurate as possible, Geant4 physics package classes for ions, hadrons, gamma particles, optical physics, and elastic collisions were included along with standard packages. 14 runs were performed with 10,000 neutron events in each, at following neutron energy levels: 1 eV, 2 eV, 5eV, 10 eV, 100 eV, 200 eV, 500 eV, 1 MeV, 2 MeV, 5 MeV, 10 MeV, 50 MeV, 100 MeV, 200 MeV.

The sampling calorimeter-style detector [13,14,15,16], consisting of layers of a thermalizer [17,18,19,20] and an active medium of scintillating borosilicate glass, should be effective for fast neutron detection (Figure 5). The ^{10}B- and natural Boron-based scintillating glasses were alternated in each layer so as to maximize the neutron-gamma discrimination. In this design the generated light is read from the edge of the active material using PMTs, similar to our previous calorimeter studies[13,14]. Figure 5 illustrates the sampling simulation setup: A 5 MeV neutron is shot at 15 layers of polyethylene-scintillating borosilicate glass pairs. The green paths are the scintillating photons produced from Cerium ions. In the final model the individual layers will be wrapped with reflective materials such as TYVEK or Mylar to reflect UV photons; in the simulated model the glasses are wrapped with 95% reflective material. Some of the generated photons reflect at the glass-PMT boundary and some of them continue through into the PMT for detection.

For geometry optimization studies, the active media (scintillating glass) was first held at 5 mm thickness and the thermalizer layer thickness was varied. Then the thermalizer was kept at a thickness of 5 mm and the scintillating glass thickness was varied. The number of captured neutrons peaks with 4 mm thick thermalizer, and decreases as the polyethylene volume equals the active volume of material. On the other hand, the number of captured neutrons versus the glass thickness plateaus at ~5 mm (Figure 6). It is logical to think that the number of captured neutrons will increase as the glass gets thicker, however for practical reasons we decided to use 5 mm thick glass samples in the simulation model. Since 10 mm thick glass was produced, it is a realistic size for a detector.

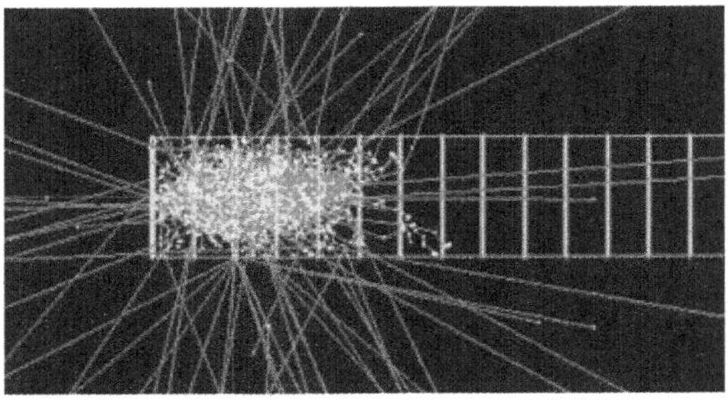

Figure 5. The Geant4 model for a sampling neutron detector with 15 layers.

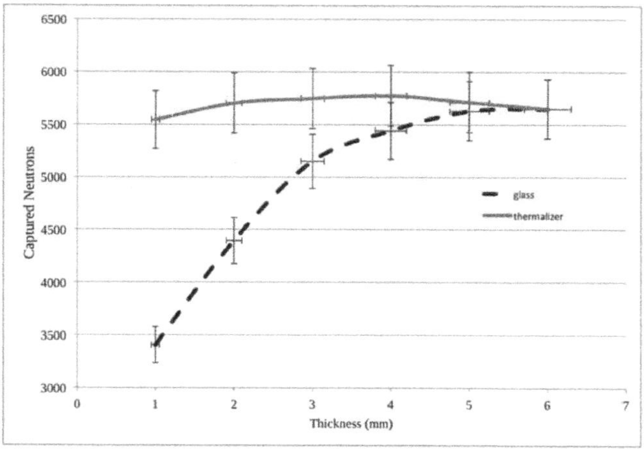

Figure 6. The glass (dashed, black line) and thermalizer (red, solid line) thickness optimization results show that the captured neutron number converges to peak values at 5 mm thicknesses.

As the energy of the neutron decreases, the probability of a ^{10}B nucleus capturing it increases dramatically. The simulation model showed that the number of captured neutrons at or below 5keV is nearly 10 times larger than the number captured at 100keV, necessitating the use of a Hydrogen-rich thermalizer (Figure 7). The detector's performance simulations also verified the produced alpha and gamma energies via neutron capture. The theoretical values of alpha energies are 1.78 MeV (6 % of the time) and 1.47 MeV (94% of the time), and can easily be observed in the simulated alpha particle energy distribution (Figure 8). Meanwhile, the produced gamma energy distribution yields a peak at 0.4 MeV as expected (Figure 9). Both of these values are compared to the captured neutron energies, and no correlation between alpha/gamma energies versus the captured neutron energy were observed.

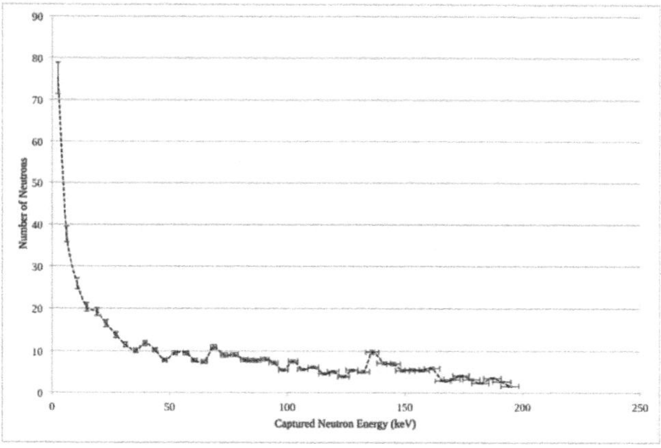

Figure 7. Geant4 simulations on our prototype detector show that the number of neutrons captured

increases drastically as the neutrons are slowed down.

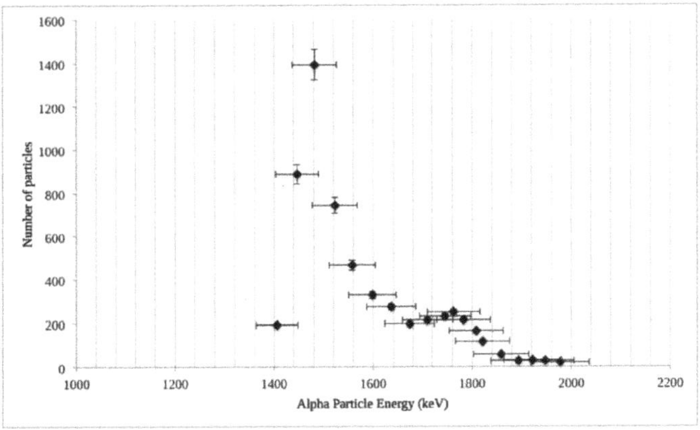

Figure 8. Geant4 simulations on our model yields the alpha particle energy peaks at 1480 keV and 1780 keV.

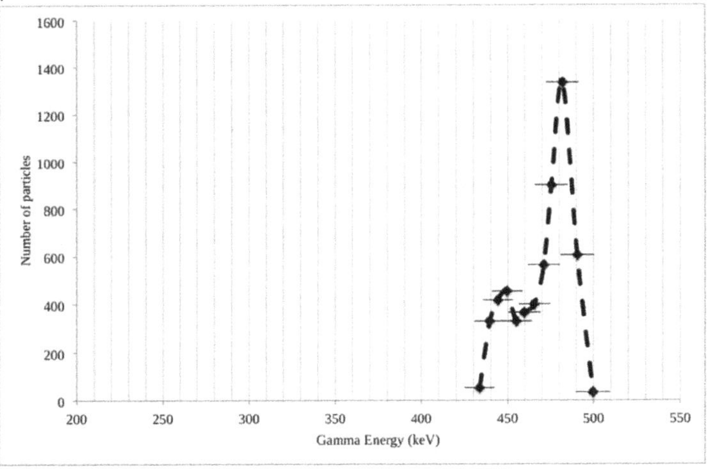

Figure 9. The gamma particle energy distribution of our simulations peaks at 480 keV as expected.

5. CONCLUSIONS

^{10}B-enriched Cerium-doped glasses can be used for neutron detectors in many areas, including homeland security. Here, the glass forming conditions for a stable, scintillating glass samples are outlined.

After pursuing high ^{10}B concentration, a relatively low content was chosen for the glass studied. The final composition used in this study is roughly similar to 1962 report by Ginther *et al.*[4], but the use of ^{10}B rather than natural Boron produced similar effects with significantly less

B_2O_3. Furthermore, the composition studied has been shown to be producible in large batches, unlike previous reports[4].

The glass samples were tested with the UV/Vis/NIR and photoluminescence instruments at Clemson University Department of Materials Science and Engineering, as well as the minimum ionizing particle (MIP) beam at CERN and Fermilab facilities. The samples are found to emit visible and UV light peaking at 390 nm. The samples were also found to be effectively transparent to the emission wavelength. The light scintillation yield was measured to be around 10k photons per MeV of energy deposited in the sample. Finally, simulated prototype detector performance and geometry optimization studies were run using Geant4 simulations. A possible prototype with 5mm glass and 4mm polyethylene layers were found to be capable of detecting a neutron capture interaction by ^{10}B atoms within the glass. A possible prototype detector would be capable of separating gamma radiation from neutron by alternating ^{10}B and natural Boron glass layers.

REFERENCES

[1]G.F. Knoll, "Radiation detection and measurements", John Wiley and Sons, Inc., New York, 2000.

[2]A. O. Hanson and J. L. McKibben, "A Neutron Detector Having Uniform Sensitivity from 10 KeV to 3 MeV", Phys. Rev. 72 673, 1947.

[3]Bollinger, L. M., Thomas, G. E., & Ginther, R. J. “Glass scintillators for neutron detection”, Review of Scientific Instruments, 30 1135-1136, 1959.

[4]Bollinger, L. M., Thomas, G. E., & Ginther, R. J. “Neutron detection with glass scintillators”, Nuclear Instruments and Methods, 17 97-116, 1962.

[5]Kojima, T., Katagiri, M., Tsutsui, N., Imai, K., Matsubayashi, M., & Sakasai, K. (2004). “Neutron scintillators with high detection efficiency” Nuclear Instruments and Methods in Physics Research Section A: Accelerators, Spectrometers, Detectors and Associated Equipment, 529 325-328, 2004.

[6]Ishii, M., Kuwano, Y., Asai, T., Asaba, S., Kawamura, M., Senguttuvan, N., ... & Shimizu, H. M. “Boron based oxide scintillation glass for neutron detection”, Nuclear Instruments and Methods in Physics Research Section A: Accelerators, Spectrometers, Detectors and Associated Equipment, 537 282-285, 2005.

[7]Bishay, A. M. “Glass Scintillator for Neutron Detection”, Journal of the American Ceramic Society, 44 231-233, 1961.

[8]S. Feller, J. Kottke, J.Welter, S. Nijhawan, R. Boekenhauer, H. Zhang, D. Feil, C. Paramswar, K. Budhwani, M.Affatigato, G.Bhasin, S. Bhowmik, J. MacKenzie, M. Royle, S. Kambeyanda, P. Pandikuthira, and M.Sharma, "Physical Properties of Alkali Borosilicate Glasses", Proceedings of the Second International Conference on Borate Glasses, Crystals and Melts (Ed. A.C. Wright, S.A. Feller, A.C.Hannon) Society of Glass Technology:Sheffield,UK) pp.246-253. 1997.

[9]K. Budhwani and S. Feller. "A Density Model for the Lithium, Sodium, and Potassium Borosilicate Glass Systems", Physics and Chemistry of Glasses, 36(4), 183-190, 1995.

[10]S. W. Martin, J.W. MacKenzie, Bhatnagar, S. Bhowmik, S. Feller, and M.L. Royle, "^{29}Si-NMR Study of the Short Range Order in Sodium Borosilicate Glasses", Physics and Chemistry of

Glasses, 36(2), 82-88, 1995.

[11]"Geant4—a simulation toolkit", Nuclear Instruments and Methods in Physics Research A 506, 250-303, 2003.

[12]"Geant4 developments and applications", IEEE Transactions on Nuclear Science 53 No. 1, 270-278, 2006.

[13]F. Duru, U. Akgun, E. A. Albayrak, A. S. Ayan, P. Bruecken, E. Norbeck, J. Olson, Y. Onel, D. Winn, L. Cremaldi, "CMS Hadronic EndCap Calorimeter Upgrade Studies for SLHC - Cerenkov Light Collection from Quartz Plates", IEEE Transactions on Nuclear Science, Vol 55, Issue 2, 734-740, 2008.

[14]U. Akgun, E.A. Albayrak, G. Aydin, B. Bilki, P. Bruecken, K. Cankocak, W. Clarida, L . Cremaldi, F. Duru, A. Moeller, A. Mestvirishvili, Y. Onel, F. Ozok, J. Parsons, D. Sanders, N. Sonmez, J. Wetzel, D. Winn, T. Yetkin, "CMS Hadronic EndCap Calorimeter Upgrade Studies for SLHC Pterphenyl Deposited Quartz Plate Calorimeter Prototype", IEEE Transactions on Nuclear Science, Volume 57, Issue 2, 754-759, 2010.

[15]U. Akgun, E.A. Albayrak, G. Aydin, W. Clarida, F. Duru, V. Khristenko, A. Moeller, J. Neuhaus, Y. Onel, J. Wetzel, D. Winn, T. Yetkin, "Quartz Plate Calorimeter Prototype With Wavelength Shifting Fibers", Journal of Instrumentation, JINST_002P_0412, 2012.

[16]G.Bhasin, A. Bhatnagar, S. Bhowmik, C. Stehle, J. MacKenzie, M. Affatigato, S. Feller, and S. Martin, "Short Range Order Structure in Sodium Borosilicate Glasses Obtained from Deconvolution of ^{29}Si MAS NMR Spectra", Physics and Chemistry of Glasses 39(5) 269-274, 1998.

[17]B.G. Parkinson, D. Holland, M.E. Smith, C. Larson, J. Doerr, M. Affatigato, S.A. Feller, A.P. Howes, and C.R. Scales, "Quantitative measurement of Q3 species in silicate and borosilicate glasses using Raman spectroscopy", J. Non-Cryst. Solids 354 1936-1942, 2008.

[18]U. Akgun, B. Bilki, P. Debbins, J. Neuhaus, E. Norbeck, Y. Onel, L. Sensoy, J.J. Sunderland, D.Winn, "Boron and Thermal Neutron Interactions on Borosilica Window Photomultiplier Tubes", Journal of Instrumentation, JINST 5 P08005, 2010.

[19]M. Jung, J. Morel, G. Nurdin, C. Teissier, and P. Siffert, "Thermalization Patterns for Broad Neutron Energy Range Real-Time Semiconductor Personal Dosimetry", IEEE Trans. Nucl. Sci. 49 2541, 2002.

[20]M. Irfan and S. A. Husain, "Thermalization of 14 MeV Neutrons and Cadmium Ratios for Iodine and Silver", Int. Jour. Appl. Radiat. Isot. 33 637, 1982.

ENGINEERING APPROACH TO IMPROVE THE SOLID STATE LIGHTING CHARACTERISTICS WITH TRANSLUCENT POLY CRYSTALLINE ALUMINA

Keiji Matsuhiro[1], Keiichiro Watanabe[1], Tsuneaki Ohashi[1], Tomokatsu Hayakawa[2]
[1]NGK Insulators, Ltd. 2-56 Suda, Nagoya 467-8530, Japan
[2]Department of Frontier Materials, Nagoya Institute of Technology, Gokiso, Showa, Nagoya 466-8555, Japan

ABSTRACT

Translucent poly crystalline alumina (TL-PCA) is the one of the earliest modern ceramic components designed for lighting applications. In order to realize the brightness and saving energy advantages better than quarts tubes, optical characters with high mechano-chemical properties in the mass productive manufacturing processes, many aspects of the ceramics sintering phenomena had been investigated and successfully performed. Driven by many pioneering persons' efforts, the quantity of production volume of TL –PCA tubes for high pressure sodium lamps started from less than 1 million pcs in 1970 growing up to 100 million pcs in 2010 including the tubes for ceramic metal halide lamps. Although an obvious technology revolution has occurred from the discharge lamps to the solid state lamps such as LED, its speed is slowing down in the last several years, however, the diversified efforts are enhancing a new application of TL-PCA products in the LED lighting field. We will here review a progress of the production technologies progress in the past and how to guide it to future from the industrial engineering point of view.

1. INTRODUCTION

Translucent poly crystalline alumina (TL-PCA) was invented by Dr. Robert L. Coble of GE in 1955. [ref. 1] It had led to High Pressure Sodium (HPS) lamps, the first commercial consumer product using the modern mass production ceramic component, in 1964. Dr. W. D. Kingery of MIT had worked together with Dr. Coble closely to achieve the wide range of the academic tasks involved throughout the densification process and microstructure controlling methodology. Not only the alumina material investigation to realize the translucency as high as possible for the light source envelop but also the chemical corrosion stability, the productive forming process technology, brightness improvement, lamp life management etc. were investigated. Combing all together, GE Lucalox™ HPS lamps was introduced into the commercial market. US patent number US 3026177, US3026210 – Transparent Alumina and Method of Preparation-1962-03-20 is the most innovative and identical patent for the people working in this field. Alumina is the most well-known ceramic at that time and even now in the electronics application field, mechanical application field, chemical equipment field and so on. In the optical application field, on contrary, the situation is relatively different, and sapphire crystal is sitting down at the main stream.

2. COMMERCIALIZATION

TL-PCA is the very unique material due to the flexible forming capability with optical translucency. Depending upon these characteristics, TL-PCA could sit down at main stream of HPS lamp key component. A variety of commercial TL-PCA tube products are shown in Photo 1. The record of the production volume is hard to find until GE's early time HPS lamp in 1965 and it was less than 1 million pcs in 1970. However, in my estimation, the quantity in 1980 had become 10 million pcs and then had increased to 100 million pcs in 2010. Within 30 years, the production volume expanded 10 times bigger.

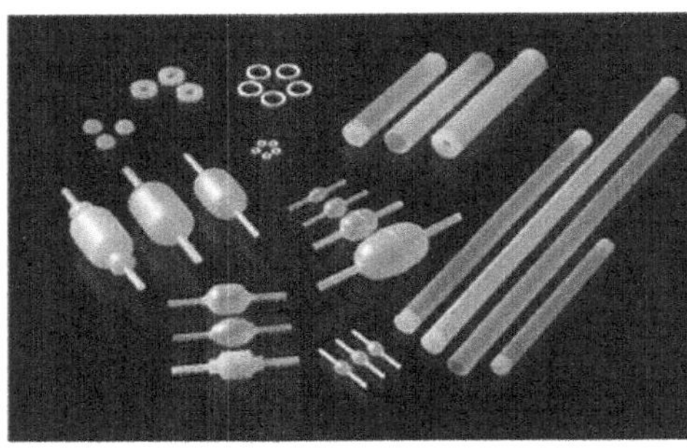

Photo 1 TL-PCA tubes for HPS and Metal Halide lamps

3. ARGUMENTS

Photo 2 shows the famous text book "INTRODUCTION TO CERAMICS" Edited by Dr. W. D. Kingery, Dr. H. K. Bowen and Dr. D. R. Uhlmann. The first edition was published in 1960. [ref. 2] This was the Bible for the research engineers and students. I was also one of students at that time. Once I had an opportunity to have a lecture done by Dr. Kingery during his visiting stay at Kyushu Univ. You can easily understand that the textbook was published more than 50 years ago but that the theoretical standpoint is perfect even now. I am so happy to have a Dr. Kingery's signature on my textbook.

Photo 2 "INTRODUCTION TO CERAMICS" [ref. 2]

The main contents of the textbook are densification mechanism, mass transportation path and driving force, densification rate prediction and grain micro structure optimization. For students, this is perfect enough to understand the ceramics densification or sintering phenomena. For the industrial field engineers, however, this is insufficient to manage the whole production process. Figure 1 shows NGK's typical TL-PCA properties. At the same time to make the products, you have to see that the material must be built to achieve the necessary functions and durability in an appropriate cost. Figures 2-7 indicates how the real components look like and how to fabricate it. Particularly, a demand for complexity and accuracy of the products dimension, which become finer and finer, has been increased very rapidly and at the same time the cost pressure with shorter process lead time has been the front line menu. In order to deal with these tasks I am seeing that a more elegant solution approach must be found out.

		G-30, G-40
Color		Translucent white
Crystal structure		α-Al_2O_3
Purity	%	99.9
Average grain size	10^{-6}m	35
Total light transmittance	%	96^{*1}
Density	g/cm^3	3.99
Water absorption	%	< 0.1
4-point flexural strength	MPa	300^{*2}
Fracture toughness K_{IC} (SEPB method)	MPa·m$^{1/2}$	4.2
Young's modulus	GPa	410
Poisson ratio		0.24
Vickers hardness	Hv	1850
Thermal expansion coefficient (40~800°C)	K^{-1}	8.1×10^{-6}
Thermal conductivity	W/(m·K)	33
Specific heat	J/(kg·K)	790
Volume resistivity	Ohm·cm	1×10^{16}
Withstand voltage	kV/mm	20
Dielectric constant		10

*1:Inner diameter=8mm, Thickness=0.75mm, Total length=105mm
*2:JIS R 1601

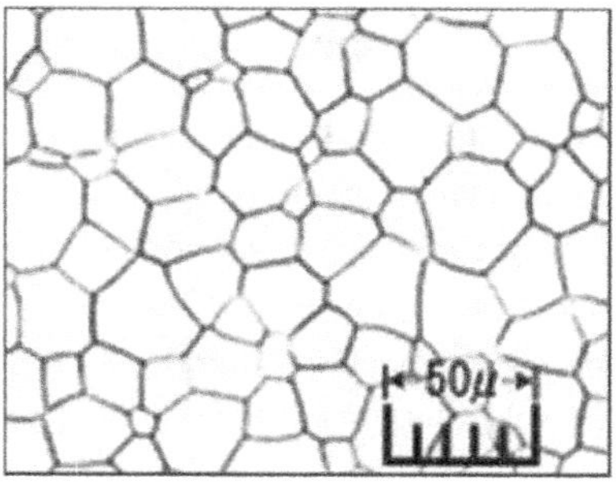

Fig. 1 NGK TL-PCA nominal properties and G-30 microstructure

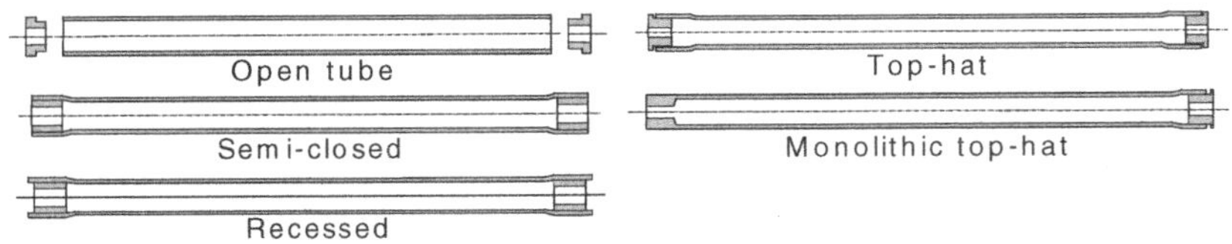

Fig. 2 Assembling structures for TL-PCA component. Straight tubes are procuded by an extrusion method. The design variation is for demonstration purpose only.

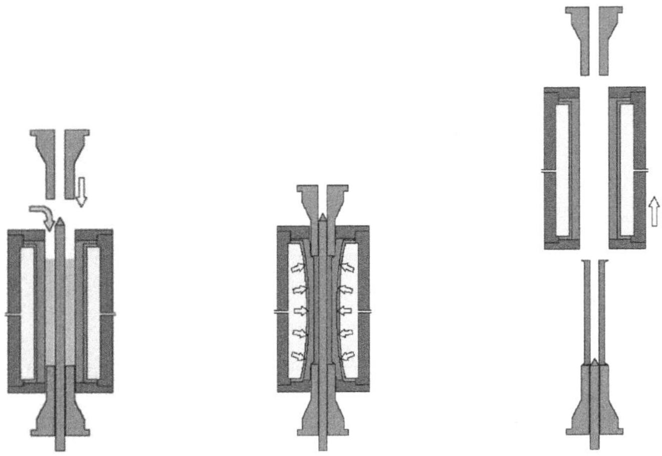

Step 1 Powder filling Step 2 Isostatic pressing Step 3 Demolding

Fig. 3 TL-PCA components forming by Dry-Bag-Isostatic-Pressing

The key elements of development tasks are to achieve near net shape simultaneously with the densification control, the translucency management, the lower energy consumption and the quicker process cycle in surrounded mild atmosphere shown in Figs. 2 and 3. Today's standard process is isostatic press forming, multi piece set up, hydrogen atmosphere, 1800 centigrade, several hours process cycle and so on. We are facing the technology limitation to break through. The ceramic metal halide lamp envelops are much more complicated, as shown in Figs.4-7. Not only the ceramics knowledge but also wide range variety of the engineering knowledges must be consolidated together.

You can easily understand that the progress of translucent poly crystalline alumina and HPS lamp technology was significant comparing to age of beginning. Now we are going to enter the key task to improve translucency of TL-PCA.

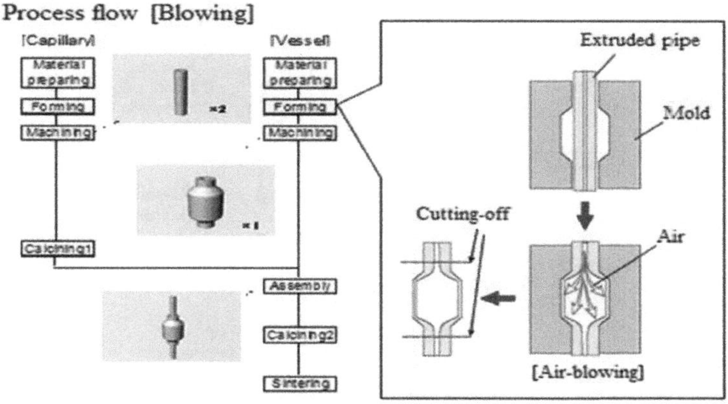

Fig. 4 Blow forming principle

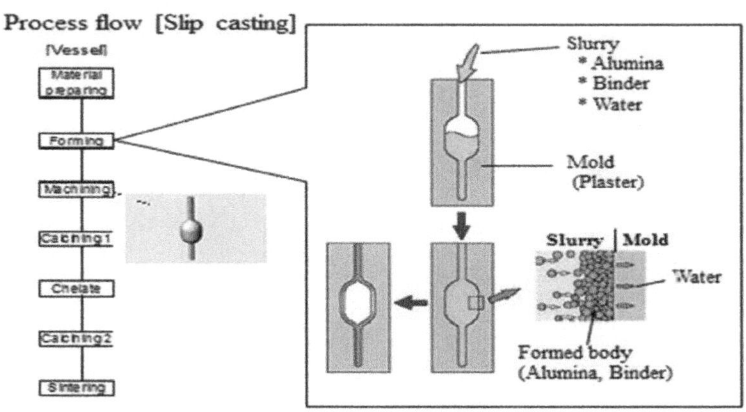

Fig. 5 Slip cast forming principle

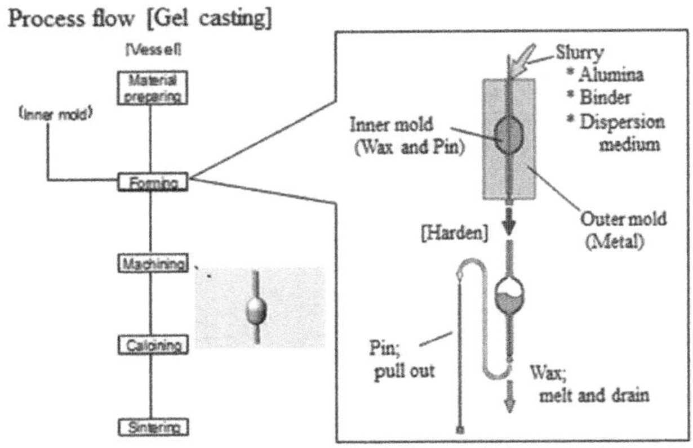

Fig. 6 Gel cast forming principle

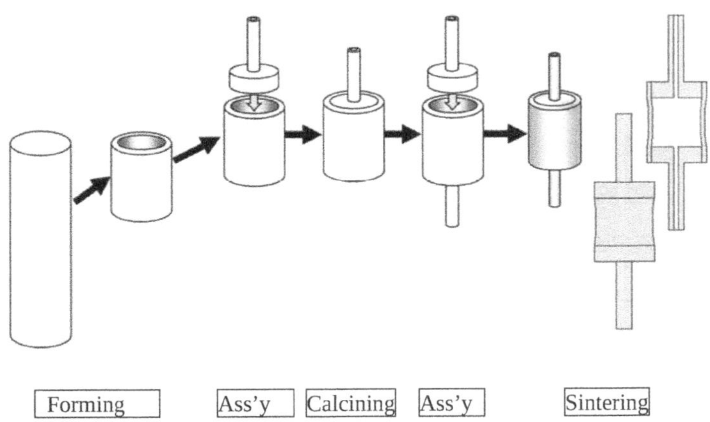

Fig. 7 Assemble individual parts to unify

In general, TL-PCA is defined as polycrystalline alumina which can transmit light. The main advantages are translucency, high mechanical strength, high toughness, high corrosion resistance and cost performance, while there is a drawback that real in-line transmission (RIT) is still lower than the corresponding single crystal alumina (sapphire). We have to solve the challenging task. The causes of the lower RIT are light scatterings due to pores, impurities and grain boundary. Pores and impurities factors can be controlled by the sintering conditions, temperature, atmosphere, etc. On the other hand, grain boundary is an unavoidable factor, because the crystal structure of alumina belongs to corundum-type hexagonal system, and exhibits birefringence with optical anisotropy. Grain-boundary scattering coefficient γ_{gb} is given by

$$\gamma_{gb} = N \cdot C_{sca,gb} = \left(V_f \frac{3}{4\pi G^3}\right) \cdot \left(\frac{8\pi^3 G^4}{\lambda^2}\left(\frac{\Delta n}{n}\right)^2\right) = V_f \frac{6\pi^2 G}{\lambda^2}\left(\frac{\Delta n}{n}\right)^2 \qquad (1)$$

where N is density of spheres in the model of grains (scatters), $C_{sca,gb}$ is scattering cross section of one sphere, V_f is volume fraction of scatters, grain size (radius), n is refractive index of the scatter, and Δn is refractive index difference between scatters and their surroundings.

In practice, the way to minimize the grain-boundary scattering of PCA is how to manage the grain-boundary scattering coefficient γ_{gb}, as explained in the above consideration. To realize smaller grain size G and smaller refractive index difference Δn is the place to work for a field ceramic engineer. An ideal goal might be single crystal sapphire; however, its forming flexibility is very limited comparing to ceramics. This is the one of points to drive the material development direction.

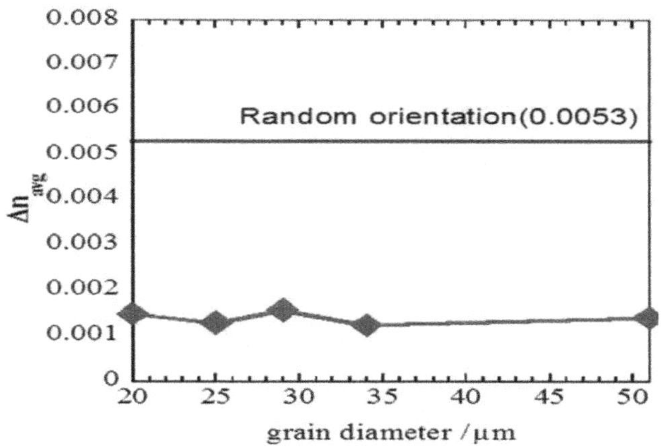

Fig. 8 Δn value verification [ref. 3] There was obtained no grainsize dependence of the averaged refractive index difference Δn_{avg}, which was found to be $\Delta n_{avg} \cong 0.0015$ for all the TL-PCA samples, smaller than a value of $\Delta n_{avg} = 0.0053$ wen grains were oriented at random.

A theoretical prediction indicated for Δn is 0.0053 (for random orientation) and the measured value is 0.00147 in average regardless the grain size of TL-PCA, as shown in Figs. 8, 9 and 10. There should be the reason to fill up this gap. Dr. Hayakawa and Dr. Hayashi et al. had investigated the gap carefully by Polarized Raman mapping method and found out why. [ref. 3-4]

All those data were collected on NGK TL-PCA. The obvious conclusion is that during densification process, neighboring individual grains had rearranged their crystallographic orientation to align for one single texture. Nearest grains had the tendency to orient their C-axis in parallel, as shown in Fig. 10. This caused the Δn smaller than the theoretical prediction on random orientation.

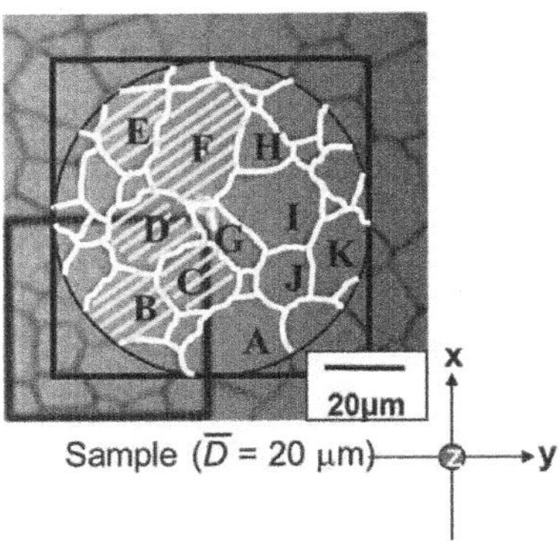

The refractive index difference Δn at each boundary.

	y polarization	z polarization
Δn_{BC}	0.00042	0.00086
Δn_{CD}	0.00435	0.00042
Δn_{DF}	0.00285	0.00020
Δn_{FE}	0.00005	0.00480

Fig. 9 Microstructure observation and Δn measurement results.[ref. 3] When an incident light advances Grain B-C-D-F-E, the light experiences Δn at each boundary. We considered as many other optical paths as possible. Eventually, the averaged measured value Δn_{avg} is found to be 0.00147.

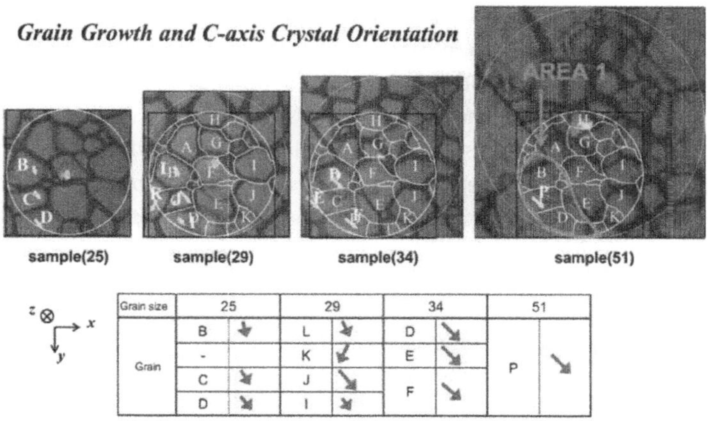

Fig. 10 Grain growth and C-axis orientation observation results [ref. 3]

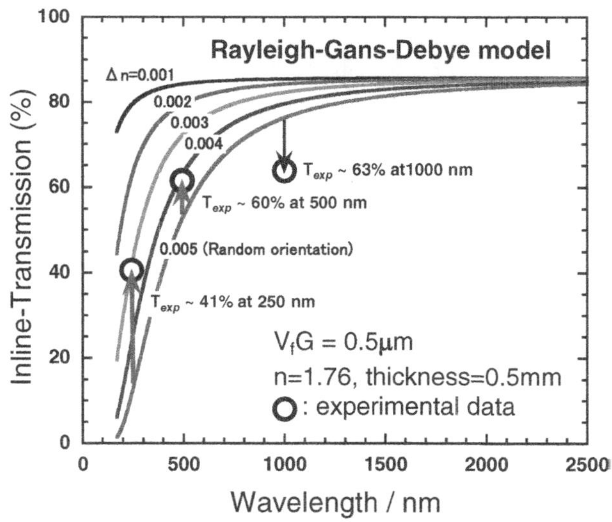

Fig. 11 Inline-Transmission result (experiment) of NGK TL-PCA and simulated curves based on Rayleigh-Gans-Debye model with various Δn [ref.3].

As shown in Fig. 11, the experimental data points at 500 nm and 250 nm are still higher than the simulated curve obtained on the assumption of random orientation (Δn = 0.005), which indicates that the actual Δn should be at least less than 0.004, taking into consideration additional losses due to impurities and tiny pores (Mie scattering). This is one of the tools for ceramics engineer to improve the translucency of PCA which has not been known before.

Focusing on the grain orientation controlling methodology throughout the production process, Figs. 12 and 13 is showing another approach done by Dr. Sakka and his colleagues. [ref.5] They use the strong magnetic field during colloidal forming, such as slip casting, gel casting, followed by sintering.

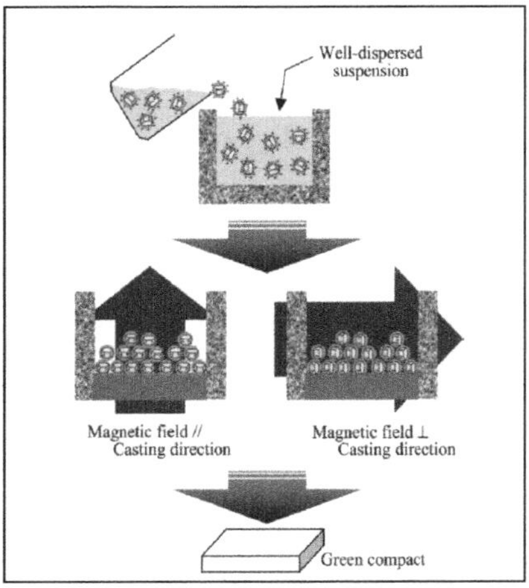

Fig. 12 Schematic diagram for fabricating textured ceramics by slip casting in a strong magnetic field [ref. 5]

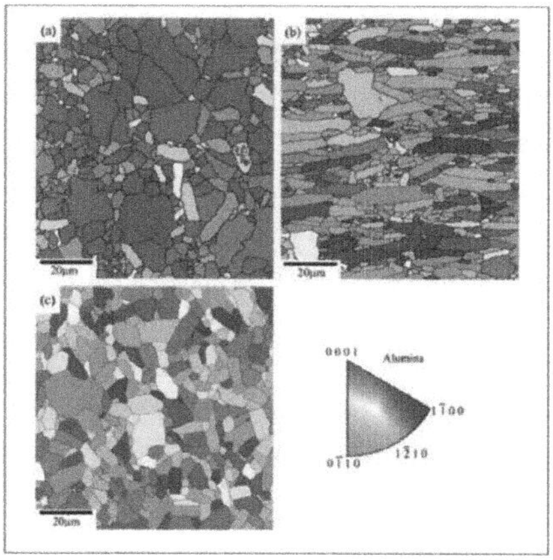

Fig. 13 EBSD map showing textured alumina by the slip casting in a strong magnetic field followed by sintering [ref. 5]

Another approach to improve the translucency of PCA is Electric Current Activated Assisted Sintering (ECAS) or Spark Plasma Sintering (SPS) also by Dr. Sakka and his colleague. [ref. 5-12] The translucent ceramics has been prepared by two-step pressure application and high pressure application. [ref. 6]

In any cases, a key technology is how to align the grain intentionally during green body forming or densification. Combining those idea and process technologies, there must be found out a much more potential way to improve translucency of PCA soon.

In early 2010s, the great technology revolution of the light emitting chip had been unveiled commercially. Light Emitting Diode, LED, had come into the market as the longer life and the higher efficiency light emitting source for the solid state light. Semiconductor wafer process is the fundamental technology against the discharge lamp technology. This is the great chance for TL-PCA to diversify the advantages for the new application. Figure 15 indicates several examples of the potential advantageous function of TL-PCA on LED.

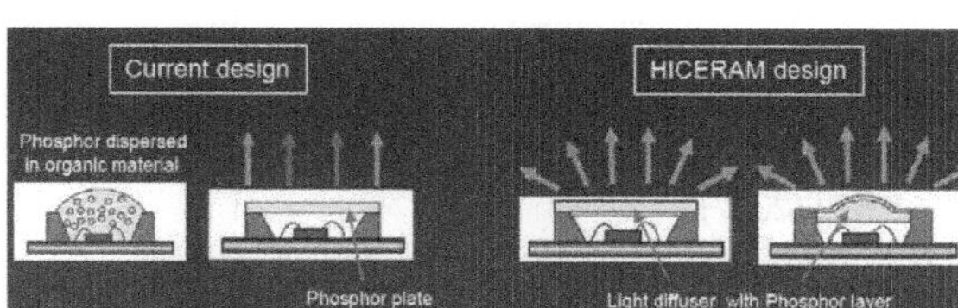

Fig. 14 Example of possible associations of TL-PCA and LED

4. CONCLUSIONS

Translucent alumina ceramics, invented by Dr. R. L. Coble in 1955, has been extended its identity for the high pressure sodium lamps and the ceramic metal halide lamps with the improving translucency, the shape flexibility, the design capability, and the multi components has been being built in higher reliability and mass production scalability has intensively been developed by academy scientists and field engineers. Due to these piled-up ideas and knowledge behind, even though the lighting technology had been switched significantly from the discharge lamp to the solid state lamp, the material and industry are able to continuously encounter the business opportunity. This review is focusing on lamp application only, however, there are much more promising applications waiting for the translucent alumina components. Authors believe the bright future on the translucent alumina with the further efforts to improve the quality and the cost.

ACKNOWLEDGEMENTS

Authors are grateful to Dr. Yoshio Sakka and his co-working research team for providing the papers and many of the kind advices.

REFERENCES

[1]R.L. Coble, Am. Ceram. Soc. Bull., 38(10),507(1959)

[2]W.D.Kingery, H.K.Bowen, D.R.Uhlmann, "Introduction to Ceramics" John Wiley&Sons, Inc. 1960, (ISBN 0-471-47860-1)

[3]T.Hayakawa, S.Hayashi, S.Miyazawa, T.Ohashi, K.Watanabe, "C-axis orientation and optical transmission properties of translucent polycrystalline alumina ceramics", *Light Sources 2012*, pp.131-132, ed. By T.Devonshire and G.Zissis, FAST-LS (ISBN 978-0-9555445-4-5).

[4]T.Hayakawa, S. Hayashi, .S. Miyazawa, T. Ohashi, K. Watanabe, "Polarized Raman mapping method for estimation of refractive index differences on grain-boundaries and in-line transmission properties of polycrystalline alumina", *Opt. Mater. Exp.* 2 (7) 914-919 (2012).

[5]T. S. Suzuki, T. Uchikoshi, Y. Sakka, "Control of texture in alumina by colloidal processing in a strong magnetic field", Science and Technology of Advanced Materials 7, 356-364 (2006)

[6]B. N. Kim, H. Hiraga, K. Morita, H. Yoshida, "Spark plasma sintering of transparent alumina", Scripta Materialia, 607-610,(2007)

[7]B. N. Kim, K.Hiraga, K. Morita, H. Yoshida, "Effect of heating rate on microstructure and

transparency of spark-plasma-sintered alumina", J. Euro. Ceram. Soc., 323-327(2009)
[8]B. N. Kim, K.Hiraga, K. Morita, H. Yoshida, T. Miyazaki, Y. Kagawa, "Microstructure and properties of transparent alumina", Acta. Materialia, 57,1319-1326(2009)
[9]B. N. Kim, K. Hiraga, K. Morita, H. Yoshida, Y. Kagawa, "Light scattering in MgO-doped alumina fabricated by spark plasma sintering", Acta. Materialia, 58, 4527-4535(2010)
[10]S. Grasso, C. Hu, G. Maizza, B. N. Kim, Y. Sakka, "Effects of Pressure Application Method on Transparency of Spark Plasma Sintered Alumina", J. Am. Ceram. Soc., 94[5], 1405-1409(2011)
[11]S. Grasso, H. Yoshida, H. Porwal, Y. Sakka, M. Reece, "Highly transparent α-alumina obtained by low cost high pressure SPS", Ceramics International, 39, 3243-3248(2013)
B. N. Kim, K. Hiraga, S. Grasso, K. Morita, H. Yoshida, H. Zhang, Y. Sakka, "High-pressure spark plasma sintering of MgO-doped transparent alumina", J. Cera. Soc. Japan, 120[3], 116-118(2012)
[12]B.N.Kim, K. Hiraga, K. Morita, H. Yoshida, Y. J. Park, Y. Sakka, "Dynamic grain growth during low-temperature spark plasma sintering of alumina", Scripta Materialia, 80, 29-32(2014)

SINGLE CRYSTAL FIBERS OF CLADDED DOPED-YAG FOR HIGH POWER LASER AND AMPLIFIER APPLICATIONS

E. Gebremichael, B. Ponting, R. Magana, and G. Maxwell

Shasta Crystals Inc.
San Francisco, CA, USA

ABSTRACT

Single crystal fibers are an intermediate between laser crystals and doped glass fibers. They have the advantages of both guiding laser light and matching the efficiencies found in bulk crystals, which is making them ideal candidates for high-power laser and fiber laser applications. This work focuses on the growth of a flexible fiber with a core of dopant (Er, Nd, Yb, etc…) and a polycrystalline clad of either YAG or alumina that will exhibit good wave guiding properties. Direct growth or a combination of growth and cladding experiments are described. Scattering loss measurements at visible wavelengths along with dopant profile characterization are also presented. Damage threshold results for these fibers are also presented. For single pass amplification, a single pass linear gain of 7.4 was obtained for 29 nJ pulses of 5 ns duration at 1 MHz repetition rate, at this gain, no parasitic lasing was observed, indicating that 7.4 is not the limit. We also obtained a nearly diffraction limited amplified beam, and a polarization ratio exceeding 250:1 at all pump power levels. In addition, we observed no damage from multiple days of operation with 80 W of the pump power absorbed in a 3 cm long Ø0.5 mm crystal fiber. These results confirm the potential for single crystal fibers to overcome the limitations of the glass fibers commonly used in fiber lasers, making them prime candidates for high power compact fiber lasers and amplifiers.

INTRODUCTION

Single-crystal fibers of yttrium aluminum garnet (YAG,$Y_3Al_5O_{12}$) provide a potential pathway to fiber lasers with higher output power. Compared with amorphous silica fibers, single-crystal YAG fibers offer higher thermal conductivity, higher Stimulated Brillouin Scattering (SBS) thresholds, higher melting temperature, and higher doping concentration and higher doping concentrations[1], as well as excellent environmental stability. Table I compares the thermal, physical, and optical properties of amorphous silica fibers and single crystal YAG fibers[2].

Table 1. Properties of Amorphous Silica Fibers and Single-Crystal YAG Fibers (T_g = glass softening temperature; T_m = melting temperature).

	T_g or T_m	Thermal Conductivity (W/m/K)	Hardness (kg/mm^2)	Theoretical Strength (GPa)	Rare Earth Dopant Concentration (%)	Refractive Index Change of Core with Temp (K^{-1})	Brillouin Gain Coefficient (m/W)
Silica Glass	~1000 (T_g)	~1	500	14.6	≤1	11.8 x 10^{-6}	5 x 10^{-11}
YAG Single Crystal	1950 (T_m)	~10	1350	56.0	~10	9 x 10^{-6}	<0.01 x 10^{-11}
Advantage of YAG Crystals Over Glass	2x	10x	>2x	>2x	10x	1.3	>100x

Equation 1 shows a formal expression for the power limit of a fiber laser due to SBS and thermal lensing[3]:

$$P_{Limit}^{SBS-TL} = \pi\lambda \sqrt{\frac{\eta_{laser} \cdot k \cdot \Gamma^2 \cdot 17 \cdot \ln(G)}{2 \cdot \eta_{heat} \cdot \frac{dn}{dT} \cdot g_B(\Delta\nu)}} \quad (1)$$

where λ is the laser wavelength, G is the gain, η_{laser} is the optical-to-optical efficiency, η_{heat} is the fraction of the pump power converted to heat, Γ is the mode field-core overlap, dn/dT is the refractive index change of core with temperature, k is the thermal conductivity, and $g_B(\Delta\nu)$ is the peak Brillouin gain coefficient. By substituting the values from Table 1 into Equation 1, we find that the power limit of a laser with a single-crystal YAG fiber gain medium is about 50× higher than that of a laser with an amorphous silica fiber gain medium.They can absorb less at longer wavelengths and generate, in theory, 50 times more power than a silica fiber laser.

Several kilowatts of CW output power have been demonstrated in ytterbium doped glass fiber, with excellent efficiency[4]. In fiber lasers, the pump is guided in a medium that is about 100 µm in diameter and the laser signal is guided in a single mode core (about 10 µm). The low absorption of the pump is spread over a long length of fiber (around 1 m), which makes the heat load better dissipated than in a bulk crystal where the pump is absorbed over a few millimeters only. However, fiber lasers exhibit limited performance in pulsed regime because high peak powers confined in a small core induce nonlinear effects that affect the quality of the laser beam.

EXPERIMENTAL

We prepared single crystal doped YAG fibers from ceramic feeds. Once doped, YAG fibers were grown by the Laser Heated Pedestal Growth (LHPG) method and then annealed in a high temperature oven. In an attempt to decrease scattering, some fibers were cladded with pure YAG and the scattering results before and after cladding were compared. We also analyzed the dopant distribution in the single crystal Nd:YAG fiber.

Ceramic feed production

The initial crystal growth was carried out using a ceramic feed stock, the preparation of which is outlined as follows. Powders of 99.99% of yttrium and aluminum oxide (American Elements) were combined with various dopants (neodymium, erbium, or ytterbium) with concentrations between 1 % atomic and 5 %atomic and mixed in stoichiometric amount in a mixer mill (8000D SEPX sample prep) for 30 minutes. The resulting mixture was then cold-pressed into the form of a pellet, using seven tons of pressure in a one inch die press. The pellet was sintered in air, inside a clean ceramic crucible at 1200°C for 24 hours. After sintering, the pellet was diced into 0.9×0.9 mm diameter square feed rods, between 20-24mm in length.

Crystal fiber growth

Single-crystal doped YAG fibers were grown by the Laser Heated Pedestal Growth (LHPG) method. Figure 1 illustrates this method, which Professor Robert S. Feigelson of Stanford University invented in 1984 [5].

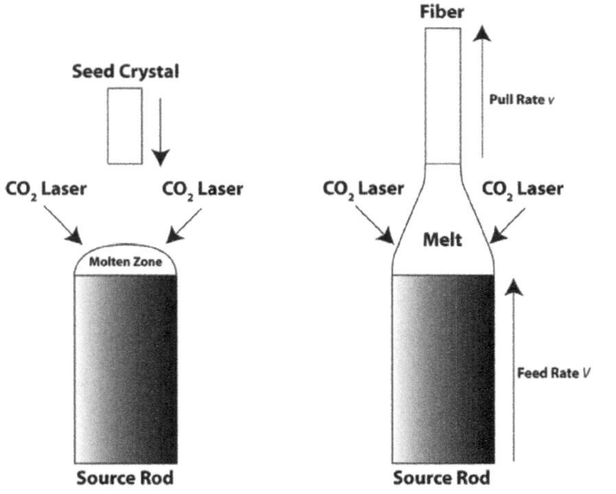

Figure 1. LHPG method.

A ceramic YAG feed rod with the desired dopant and dopant level is chosen and mounted in the LHPG growth machine. The tip of the feed rod is melted by a focused beam of infrared energy produced by a CO_2 laser. A mechanical apparatus lowers a seed crystal with the desired orientation into this molten zone. Once the seed is wetted, it is pulled upwards, drawing the molten material up with it forming the new crystal as it freezes. The melt is replenished by translation of the feed rod up into the melt as it is drawn out. The ratio between the seed pull rate, the initial diameter of the feed and the rate it is fed into the melt, determine the final diameter of the grown crystal. Typical reduction ratios are on the order of 2-3:1. The amount of power required to melt the initial ceramic feed rod is typically around 10-15W in our system. Subsequent diameter reductions of the grown crystal require proportionally less power, with the final reduction requiring less than one watt of power. The rate at which the crystal is grown is typically 1-2mm/min for the growth of 500-1000μm diameter crystals. The final fiber, 30-120μm in diameter, is grown at 3-5mm/min. Typical fibers are grown from 10-90cm in length in this way. The crystal fibers become more flexible as their diameter is reduced with fibers of ~100μm in diameter having a bend radius of ~1cm and thinner fibers having correspondingly tighter bending radii.

In an attempt to control the uniformity of the crystal fiber diameter, a computer controlled feedback system was implemented. The system focuses a red laser beam onto the growing fiber and produces a series of interference fringes. These fringes are imaged onto a CCD line camera and interpreted by a software program which displays the calculated fiber diameter. In open-loop growth, or without any computer feedback control, the diameter of the fiber will fluctuate with any change in the diameter of the feed or change in laser power. The software attempts to correct for this by allowing computer control over the feed motor speed, influencing the volume of the melt and ultimately the diameter of the grown fiber. The level of computer control in the closed-loop setting is adjusted by modifying its gain settings. Figure 2 indicates that the feedback system reduces the maximum diameter fluctuations from 7% to less than 1%.

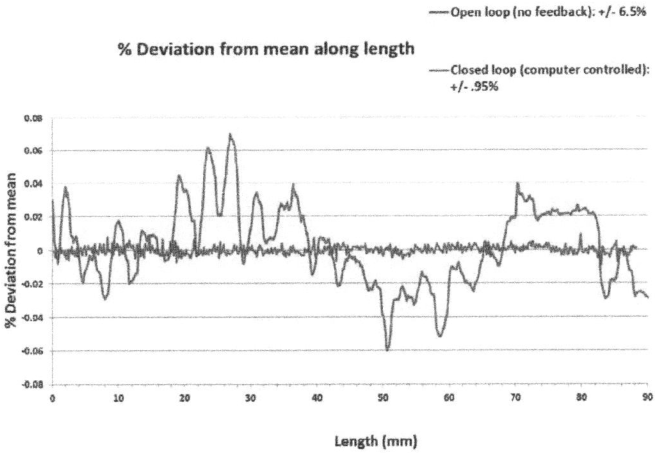

Figure 2. Impact of Diameter Control Feedback System.

Annealing and polishing of crystal fiber

Post growth, the approximately 4 inch long fibers were annealed in a high temperature oven in air. In order to maintain a clean annealing environment, the fibers were placed inside a covered alumina dish, wide enough in diameter to be at least 3-4 times the bending radius of the fiber. The oven was ramped up at a rate of 5°C/min to a temperature of 1000°C, held at this temperature for twelve hours, and then allowed to cool down until the fibers reached room temperature.
Following the annealing, the tip of each fiber was placed inside a close fitting glass capillary tube and held in place by a low melting adhesive (Crystal Bond). An ULTRAPOL End and Edge polisher was used with a series of aluminum oxide polishing pads of decreasing grit size in order to produce a laser grade polish on the fiber end faces.

Sol-gel cladding

YAG cladding

To decrease the amount of bulk scattering, sol-gel YAG fiber clad materials was explored. The material was prepared using sol-gel mechanisms that required very low reacting temperatures (under 100°C). Yttrium aluminum garnet (YAG) was made via the hydrolysis and condensation reaction of yttrium oxide (Y_2O_3) and aluminum nitrate $Al(NO_3)_3$. A modification of common sol-gel synthesis procedure was used to prepare YAG [6, 7, 8]. The reaction was carried out in acidic media to prevent flocculation and with the use of citric acid, a pacifier. Aluminum nitrate monohydrate (>98%), yttrium oxide (99.99%), 1.0 M acetic acid and citric acid monohydrate, (98%) were all purchased from Aldrich Chemical Company. Yttria was firstly dissolved in 0.2M acetic acid by stirring at 60°C for 2-3 hours while covered. The aluminum precursor was then added along with citric acid and the mixture was stirred at 80°C.
After approximately 2 hours, the increase in viscosity was evident. However the solution was not allowed to reach gelation in order to produce thin coatings that were crack free. To minimize defects, the substrates were cleaned thoroughly by soaking in acetone and then in water in an ultrasonic bath for several minutes.

The YAG substrate was dip-coated in the sol at 1-10cm/min and allowed to dry very slowly in air for 24 hrs. Once dry, the sample was heated for a dwell time of 3 hours at 500°C. To produce thicker coatings, the coating and drying process was repeated until the desired size was achieved.

Alumina cladding

Inside an Erlenmeyer flask, 11.2541 g (3.0000 x 10-3 mol) of aluminum nitrate monahydrate ($Al(NO_3)_3 .9H_2O$, >98%, Sigma Aldrich) is dissolved in 60 mL distilled water. 15.7648 g of citric acid (7.5020 x 10-2 mol, >98%) is added to this solution and the mixture is allowed to stir for 1-3 hours at 60˚C (sol is still clear at this point). See, e.g., Li, Jiang, et al. "Low temperature synthesis of ultrafine α-Al 2O3 powder by a simple aqueous sol–gel process." Ceramics International 32.5 (2006): 587-591. Then the solution is heated at 80˚C uncovered for 8+ hours until the desired viscosity is reached as determined by testing on trial fibers of dimensions similar to the fiber to be coated. At this point, a hint of yellow appears as the volume decreases and the transparent gel may now be used to coat fibers. In order to produce thin coatings that are crack free, the solution is not allowed to reach gelation.

Dip Coating

The YAG fiber was dip-coated in the sol at 1-10cm/min and allowed to dry very slowly in air for 24 hrs.

To coat a fiber thinner than 140 µm, our patent pending draw coating technology is used. A reservoir such as the one depicted in Figure 3 is used. The fiber is inserted inside a capillary tube on the lower side of the container. Once the fiber is visible in the top end of the capillary tube, a very narrow and short piece of tape maybe used to make sure the fiber does not fall through until it is secured. Next, a string with a small amount of Crystal Bond is brought very close to the top face of the crystal. The string is then aligned carefully with the fiber and approximately 1mm past the end of the fiber. The Crystal Bond at the end on the string may then be gently heated with a torch until it is soft and the string and the crystal are brought together. After the crystal bond is allowed to cool, the reservoir may be filled with the YAG sol-gel. The YAG gel is then coated on crystal by pulling up the string at a controlled rate driven by a motor. For YAG, this draw rate maybe anywhere from 18mm/min to 41mm/min depending on the viscosity of the gel. The film is then allowed to dry very slowly for approximately 23-26 hrs. When drying in air, humid conditions must be avoided. Once dry to the touch, the sample is heated to 500°C for 3 hours at a ramp rate of 1°C/min.

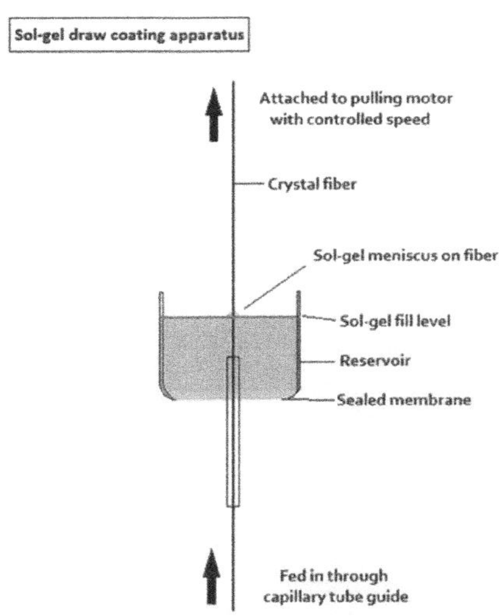

Figure 3. Patent pending Draw Coating apparatus for thin fibers.

Once dry, the sample was heated for a dwell time of 3 hours at 500°C. To produce thicker coatings, the coating and drying process was repeated until the desired size was achieved.

Scattering measurement

Bulk scattering losses inside the Yb:YAG fibers were measured by using an integrating sphere. The major components of the scattering setup are, a two inch diameter integrating sphere (Lab sphere 4P-GPs-020-sL, Spectralon), a chopper which decreases the noise during measurement at low scattering levels, and a lock in amplifier (SR810 DSP). A 532 nm (5mWatt) diode laser beam is sent through the optical chopper which also sends a reference signal to the lock-in amplifier. A small fraction of the beam is sent to a laser beam monitor detector. The beam must be focused into the sample crystal. The laser beam can pass through the crystal and the portion of the beam that is scattered by the crystal is captured and reflected by the integrating sphere and is detected by the semiconductor detector (Silicon PIN detector). The output from the detector is then sent to the lock-in amplifier.

The measurement technique involves placing a 4 inch long fiber within the integrating sphere and focusing the laser radiation in and through the sample. If both end faces of the sample are not allowed to reflect light directly into the integrating sphere, only light scattered within the sample bulk is measured by the detector. Blocking the exit port and measuring the detector's signal gives a relative measurement of the total power into the sample. Comparison of the signal obtained with the sample inside and outside of the sphere gives an absolute determination of the optical scattering losses. Chopping the laser beam allows phase-sensitive detection enhancing the signal-to-noise ratio at low scattering levels.

Energy-dispersive X-ray spectroscopy (EDS)

An EDS machine from BALAZS NanoAnalysis was used for the elemental analysis of the polished end faces of Nd:YAG crystal (diameter of 500μm, length of 8mm) Electron beam excitation is used in a scanning electron microscope (SEM). X-ray beam excitation is used in X-ray fluorescence (XRF) spectrometers. A detector (Si(Li) detector cooled with liquid nitrogen to cryogenic temperatures) was used to convert X-ray energy into voltage signals; this information is sent to a pulse processor, which measures the signals and passes them onto an analyzer for data display and analysis.

Damage Threshold Measurements

Spica Technology measured the damage threshold of a 100 micron diameter Nd:YAG and Yb:YAG single-crystal fibers that we fabricated. For the Nd:YAG fiber, they focused a CW laser at 1micron into the fiber and increased power incrementally until they observed either a change in transmission or physical damage. For the Yb:YAG fibers, they measured the damage thresholds using 6ns pulses and 10ps pulses.

RESULTS AND DISCUSSION

To fabricate single-crystal YAG fibers as laser gain media, the technical challenge is to synthesize a cladded flexible fiber with a core of dopant (such as Er, Nd, or Yb) that will exhibit good wave guiding properties. LHPG offers a competitive advantage over other techniques for growing single-crystals fibers for this application. The growing crystal accepts yttrium, aluminum, and oxygen in the garnet structure and rejects dopants with segregation coefficients different than 1. The temperature differences between the molten zone, the air, the crystal, and the feed rod create four convection cells in the molten zone. In all four cells, Marangoni convection currents drive the dopant impurities to the center, creating a natural core in Nd dopant (figure 4). This statement is what the theory predicts for dopants with large ionic radius like Nd and is not borne out with every dopant such as in the case of Er.

We can grow Nd:YAG fibers with a doped core and an undoped YAG cladding fibers up to 900 mm in length. These fibers consist of a doped core 30 microns in diameter surrounded by an undoped YAG cladding with an outer diameter of 70 to 100 microns.

To grow these fibers, we prepared ceramic pellets of YAG doped with neodymium, diced each pellet to create rods with cross-sectional areas of 1 mm^2, then grew doped YAG single crystals from these rods with growth rates ranging from 0.6 to 0.9 mm per minute. During each run, we maintained a ratio of 3 between the growth rate v and the feed rate V. Under these operating conditions, 1-mm thick rods yielded crystals 0.55 mm in diameter.

EDS measurements revealed the Neodymium-doped core. We studied the dopant concentration profiles at both polished end faces of the fibers via electron dispersion spectroscopy (EDS). Figure 4(a) shows the distribution of neodymium atoms in a YAG fiber 100 mm long and 550 microns in diameter. A neodymium-doped core occupies 30% of the overall diameter of the crystal. Both end faces exhibited the same dopant concentration profile, indicating that the doped core occupies the entire length of the crystal.

We re-grew the crystal three times. The 20 μm core could be achieved in one growth, if diameter is 60 μm ratio of core to diameter of fiber is what changes with re growths. The feed/pull ratios used was kept similar. Figure 4(b) shows the distribution of Nd after the third regrowth. The diameter of the Nd-doped core shrunk to about 50 microns, or 1/7 of the total diameter. We observed the same dopant concentration profile at both end faces.

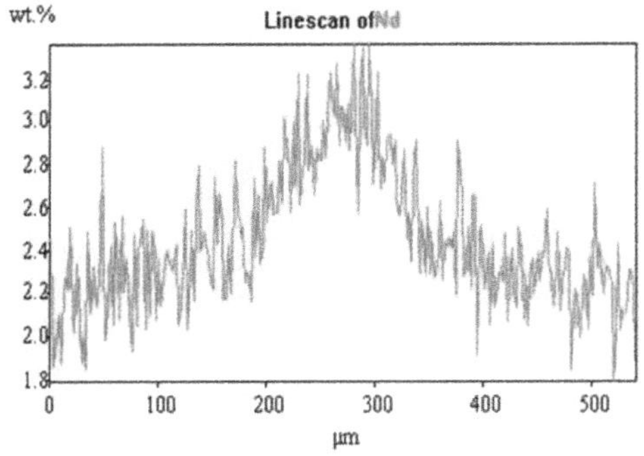

(a)

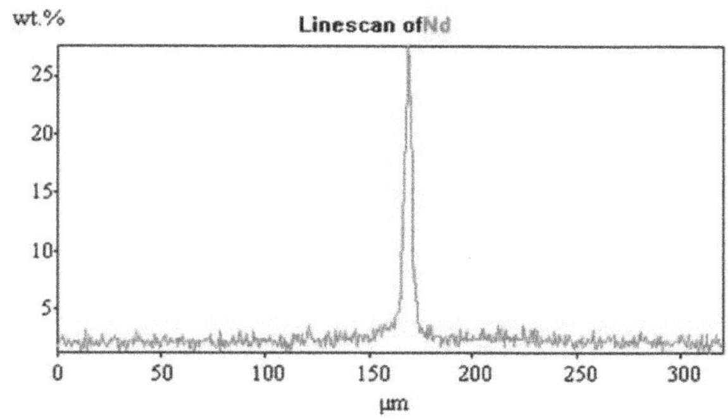

(b)

Figure 4. Line scan of Nd Concentration in Nd:YAG Single-Crystal Fiber (wt % vs. position). (a) distribution of neodymium atoms in a YAG fiber 100 mm long and 550 µm in diameter, core is 113 µm (b) distribution of Nd after the third regrowth. The diameter of the Nd-doped core shrunk to about 1/7 of diameter of fiber.

As-grown fibers satisfied the requirement for total internal reflection. We measured the refractive index of the Nd:YAG single-crystal fibers at UC Berkeley's Marvell Nanofabrication Laboratory with Ellipsometer (Sopra – 380, spot size of 75µm×150µm). The higher core refractive index (~0.7%) achieved by adding dopant and a critical angle of 80.03° for total internal reflection indicates that the natural segregation of Nd ions in the core of the single-crystal Nd:YAG fibers creates a guiding structure. The fibers will require no further external cladding to serve as gain media for lasers with Gaussian beams.

We also grew fibers from ceramic pellets doped with erbium, ytterbium, and holmium. Erbium and ytterbium are desirable because they generate eye-safe wavelengths. Ytterbium is desirable because it allows pico and femtosecond laser pulses. In each case, EDS studies revealed no core; the dopant concentration remained equal throughout the volume of the fiber. We then grew fibers from ceramic samples in which we had artificially segregated Er and Yb in the center. Again, EDS studies revealed an even distribution of dopants throughout the fiber. Our hypothesis is that LHPG cannot segregate Ho, Er, and Yb dopants within the fiber because they have smaller ionic radii than Nd.

To provide a cladding for LHPG-grown YAG fibers doped with ions such as Er, Yb, and Ho, which do not spontaneously congregate at the core, we developed a new sol-gel process to coat the as-grown fibers with polycrystalline YAG. We chose yttrium oxide (Y_2O_3) as the Y source and aluminum nitrate nonahydrate ($Al(NO_3)_3 \cdot 9H_2O$) as the Al source. X-ray diffraction studies confirmed that this novel procedure yielded single-phase cubic-face YAG (see Figure 5(a)).

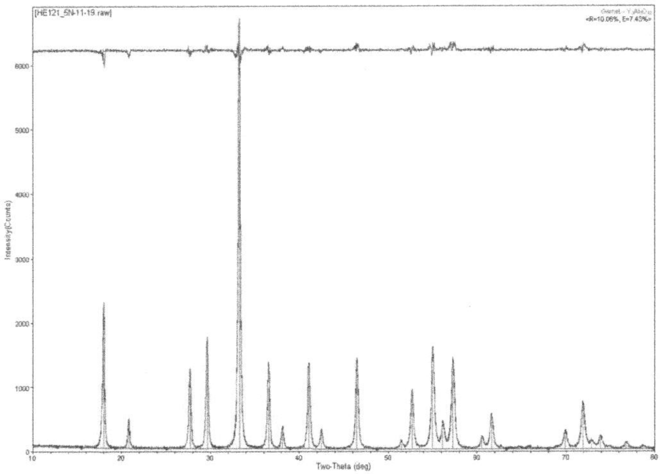

(a)

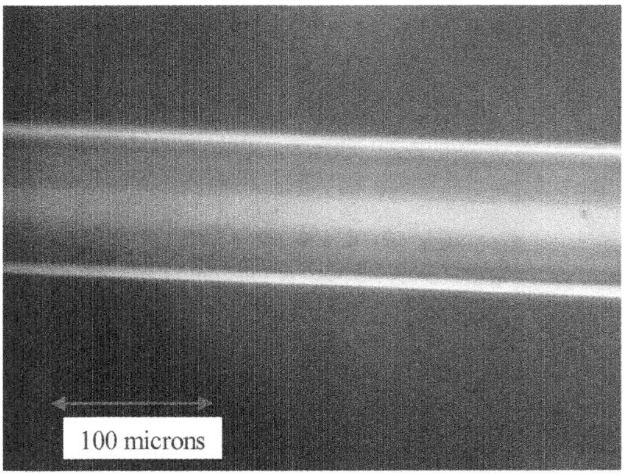

(b)

Figure 5. (a) XRD Measurements of Sol-Gel YAG Cladding Indicate Single-Phase Cubic-Face YAG Yb:YAG.(b) Yb:YAG fiber 130 µm in Diameter Cladded with 3 Sol-Gel Coatings of Polycrystalline YAG.

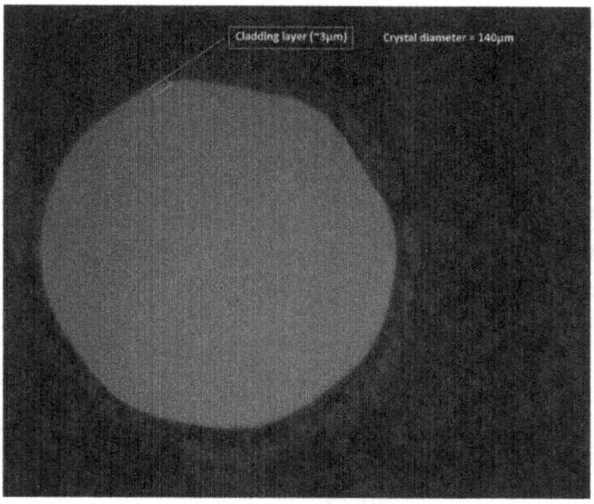

Figure 6. Undoped YAG cladding around a doped YAG core.

For such a characterization, the YAG prepared by sol-gel was allowed to gel fully. The powder was then placed in a crucible, crushed and heated at 500°C (1C/min) for a dwell of 6 hrs to safely eliminate all citrates and nitrates. Finally to produce crystalline YAG, the resulting powder was crushed in a mortar and pestle and reheated at 1000°C (8°C/min) for a dwell 3 hrs. Before the XRD measurements were made, the previous powder was again lightly ground in a mortar and pestle and pressed into a bulk sample holder with a glass slide for analysis. Data was collected by a coupled Theta:2-Theta scan on a Rigaku Ultima-III diffractometer equipped with copper x-ray tube, para focusing optics, computer-controlled variable slits, and a diffracted beam monochromator.

To our knowledge, we are the first team to fabricate a transparent, crack-free YAG fiber cladding. Figure 5(b) shows the fiber after a third coat, and figure 6 shows an undoped clad around a 140 µm core.

In order to tailor the NA of our fibers, we have invented and optimized the recipes for a variety of claddings with various indices of refraction. We can clad with pure or doped YAG, Al_2O_3, or silica.

Table 2 shows the scattering results. To gain maximum lasing efficiency, it is important to reduce scattering in single crystals fibers to levels close to the ones of silica fibers (~10 dB/km). We have found throughout our experiments that the scattering loss can be reduced by annealing the fibers at 1100˚C for 12 hours and/or by cladding them with pure YAG made by sol-gel. We have shown that we can grow the fiber by using our diameter control techniques and reach a scattering coefficient as low as 0.14 dB/m.
According to literature[9] the YAG shows the absorption due to OH related defects in both the starting source rod crystals and in the grown fiber. It is not known precisely why the losses in the visible wavelength decrease after the annealing process but it is suspected that annealing in air might somehow help to decrease the oxygen related impurities[9]. As we can see, annealing in air

and sol gel cladding as well as using the diameter control techniques have helped reducing the attenuation at 532nm for the YAG fiber.

Table 2. Yb:YAG fiber scattering results.

Seed Speed (mm / min)	Feed Speed (mm / min)	Length (mm)	Fiber Dia-meter (mm)	Scattering Coefficient (dB / m) Before Annealing	Scattering Coefficient (dB / min After Annealing)	After 2nd Layer of Sol- Gel Coating	After 3rd Layer of Sol-Gel Coating
5	0.12	121	0.12	2.80	0.58	0.47	0.044
5 (w/DC)	0.45	120	0.04	0.25	0.17		
3	0.30	120	0.11	1.50	0.75		
3 (w/DC)	0.30	98	0.12	0.14	0.14		

DC = diameter control

Damage threshold experiments (Figure 7) at a wavelength of 1.07 microns in CW regime revealed that our 100 micron diameter Nd:YAG fiber presents a damage threshold of $6MW/cm^2$. This is a very promising result, considering that the damage threshold for a bulk Nd:YAG crystal is around 1 MW/cm^2.

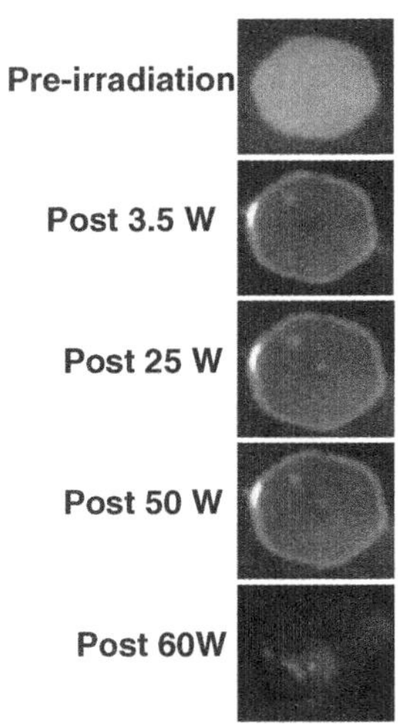

Figure 7. Nd:YAG Single-Crystal Fiber optical damage experiment.

For Yb:YAG in pulsed regimes, we found out that at 6ns pulses, the damage threshold for a 100 µm fiber was 80J/cm^2, and at 10ps pulses, the damage threshold for a 80 µm fiber was 10.2J/cm^2.

AMPLIFIER EXPERIMENTS

Small signal gain measurements were performed on an uncladded Yb:YAG rod, 500µm in diameter and 30mm long. For single pass amplification, a single pass linear gain of 7.4 was obtained for 29 nJ pulses of 5 ns duration at 1 MHz repetition rate, at this gain, no parasitic lasing was observed, indicating that 7.4 is not the limit. This may be due to a specific texture of LHPG crystals fiber surface as opposed to a mirror-like surface of the small diameter rods made by optical polishing or micro pulling down technique. We also obtained a nearly diffraction limited amplified beam, and a polarization ratio exceeding 250:1 at all pump power levesl. In addition, we observed no damage from multiple days of operation with 80 W of the pump power absorbed in a 3 cm long Ø0.5 mm crystal fiber.

These results allow us to expect the following in a Ø500 µm fiber: a very high damage threshold exceeding 30 J/cm^2 for 6 ns pulses in combination with high small signal gain and high beam quality. We can get diffraction limited output pulses with single pulse energy of 19 mJ/pulse which corresponds to peak power **exceeding 3.2 MW/cm^2**. Moreover, 2 kW of the pump power absorbed within Ø500 µm, 3 cm long fiber bring the crystal only to 1/3 of its thermal fracture limit. If we assume 40% conversion efficiency we can expect that a compact pulse amplifier with a **kW output power** can be built.

CONCLUSION

In this work, we have demonstrated the ability to grow over 900mm long and 30 microns in diameter single crystal fibers with a variety of dopants. For Nd, the maximum doping level that could be attained was 4% whereas 10% was reached for Yb and 50% for Er. We established that Nd segregates and creates a natural core. For other dopants (Er, Yb…) with smaller ionic radius, no natural core occurred and a sol-gel process to clad the fibers with pure YAG was established and optimized. The best scattering results were obtained with cladded doped fibers grown with diameter control and annealed in air. Scattering as low as 0.044dB/m was then obtained, proving that they are a very good wave guiding, low loss medium. We also have demonstrated the great potential of the LHPG – manufactured crystal fibers with their high gain, relatively simple and inexpensive manufacturing, simple amplifier geometry and excellent beam quality, thermal and mechanical properties and very high optical damage threshold.

These preliminary results are very promising and confirm the potential for single crystal fibers to overcome the limitations (Stimulated Brillouin Scattering, etc…) of the glass fibers commonly used in fiber lasers, making them prime candidates for high power compact fiber lasers and amplifiers.

REFERENCES

[1] Fair GE, Kim HJ, Lee HD, Parthawarathy TA, Keller KA, Miller ZD, Development of ceramic fibers for high-energy laser applications, *Proc. SPIE* 8039, *Laser Technology for Defense and Security VII*, 80390X (May 13, 2011)

[2] Kim W, Florea C, Baker C, Gibson D, Shaw LB, Bowman S, O-Connor S, Villalobos V, Bayya S, Aggarwal ID, and Sanghera JS, *Single Crystal Fibers for High Power Lasers, Proc. of SPIE Vol. 8547*, 85470K, 2012 doi: 10.1117/12.974848

[3] Dawson JW, Messerly MJ, Heebner, JE, Pax PH, Sridharan AK, Bullington AL, Beach RJ, Siders CS, Barty CPJ, Dubinskii M, Power *scaling analysis of fiber lasers and amplifiers based on non-silica materials, Proc. Of SPIE Vol. 7686*, 768611, 2011, doi: 10.1117/12.852393

[4] Royon R, Lhermite J, Sarger L, Cormier E., High power, continuous-wave ytterbium-doped fiber laser tunable from 976 to 1120 nm, *OPTICS EXPRESS, 21(11)*, DOI:10.1364/OE.21.013818.

[5] Fejer M, Nightingale J, Magel G and Byer R, "Laser-Heated Miniature Pedestal Growth Apparatus for Single-Crystal Optical Fibers," *Rev. Sci. Instrum. 55*, pp. 1791-17% (1984).

[6]Veith M, Sanjay M, Kareiva A, Jilavi M, Zimmer M, "*Low Temperature synthesis of nanocrystalline $Y_3Al_5O_{12}$ (YAG) and Ce-doped $Y_3Al_5O_{12}$ via different sol-gel methods,*" *Journal of Mater. Chem., 9(12)*, 3069-3079 (1999).

[7] Devi K, Choudhary R, Satsangi A. and Gupta R, "*Sol gel synthesis and characterization of nanocrystalline Yttrium Aluminum Garnet Nanopowder.*" *Journal of Def Sci, 58(4)*, 545-548 (2008).

[8] Zhou Y, Lin J, Wang S, Zhang H, "*Preparation of Y3Al5O12:Eu phosphors by citric–gel method and their luminescent properties,*" *Journal of Opt. Mater, 20(1)*, 13–20 (2002).

[9]Nubling R, Harrington J,"*Optical properties of single-crystal sapphire fibers*", *Journal of Applied Optics, 36 (24)*, 20 (1997).

SINGLE CRYSTAL GROWTH OF FERROELECTRIC $LaBGeO_5$ FOR OPTICAL FREQUENCY CONVERSION DEVICES

Shintaro Miyazawa[1,2], Mitsuyoshi Sakairi[2], Junji Hirohashi[2], Makoto Matsukura[2], Shunji Takekawa[2] and Yasunori Furukawa[2]

[1] ZAIKEN, WASEDA University, 2-8-26 Nishiwaseda, Shinjuku-ku, Tokyo 169-0051, Japan

[2] OXIDE Corp., 1747-1 Makihara, Mukawa, Hokuto, Yamanashi 408-0302, Japan

ABSTRACT

$LaBGeO_5$ (LBGO) has recently been developed as a promising ferroelectric crystal for generating UV wavelengths by periodically poled, quasi-phase-matching (PP-QPM) structures. LBGO single crystals suitable for PP-QPM devices were grown by the Czochralski-pulling. By considering the degree of supercooling of the melt, *c*-axis-oriented, relatively large LBGO single crystals with high crystallinity were successfully obtained. The optical absorption edge was verified to be around 195nm. The PP-QPM devices were successfully fabricated, and 355nm generation by mixing 1064nm and its second harmonic 532nm beams was obtained. About 500mW output power was confirmed.

INTRODUCTION

Frequency conversion devices generating UV wavelengths by second/third harmonic generations (SHG/THG) have gathered many attentions for industrial applications. There are two ways for frequency conversion, i.e., birefringence phase-matching (BPM) and quasi phase-matching (QPM). In the case of BPM, many optical nonlinear borate crystals such as β-BaB_2O_4, LiB_3O_5, $CsLiB_6O_{10}$ have been developed in practice. However, these nonlinear borate crystals are hygroscopic and show a relatively large walk-off, i.e., the generated harmonic laser beam is non-collinear with the incident beam and its beam shape is elongated to some extent as a result. This walk-off phenomenon is one of disadvantages crucial for practical optical alignments. On the other hand, so-called quasi-phase-matching (QPM) devices made of periodically poled ferroelectric crystals such as $LiNbO_3$, $LiTaO_3$ and $KTiOPO_4$ are very attractive because of no walk-off. However, most of ferroelectrics have the limitation of optical transparency in the UV region.

To overcome these issues, we have paid much attention to a ferroelectric $LaBGeO_5$ (LBGO) of stillwellite structure belonging to a trigonal system with the space group C_3^2-$P3_1$. The crystal is transparent down to ~190 nm and not hygroscopic. LBGO was early investigated for applying it to frequency-doubling lasers[1], and its dielectric properties were well investigated[2]. Recently, we reported on the fabrication of periodically poled LBGO (PP-LBGO) devices and demonstrated SHG and sum frequency generation (SFG) [3,4].

However, a detail of its bulk single crystal growth has not been reported so far. Only a few paper pointed out two major problems[5,6,7] ; a relatively high viscosity of the melt and an evaporation of B_2O_3 and/or GeO_2 during the crystal growth from melt. Especially, the tendency of supercooling, associated with its high viscosity of the melt, results in uncontrollable crystallization.

In this paper, we will focus on a single crystal growth of $LaBGeO_5$ by the Czockralski (Cz)-pulling directly from the melt and the evaluation of crystal quality. With the high quality crystals, we will demonstrate a 355 nm generation with more than 500 mW of output power by the quadratic sum frequency conversion process of a fundamental 1064 nm and its SHG-532 nm.

EXPERIMENTS

1. Crystal growth and characterization of quality

LBGO melts congruently at about 1200°C, and then the single crystals can be grown by the conventional rf-heated Cz-pulling method. The furnace configuration was already reported elsewhere[7,8]. Commercially available raw materials of 5N purity La_2O_3, H_3BO_3 and GeO_2 were mixed in a stoichiometric ratio by the conventional ball-milling. The compound calcined at elevated temperatures was charged into a Pt-crucible. The Cz growth was carried out with pulling rates of 0.5~2.5 mm/hr, crystal rotation rates of 15~30 rpm in different atmospheres (O_2, air and N_2). All of crystals described hereafter were grown along the <001> c-axis.

The grown crystals were evaluated mainly by the X-ray (Cu-Kα) diffractometry, by measuring the ferroelectric Curie temperature Tc, and by observing a conoscope figure under a polarizing optical microscope.

2. Frequency conversion experiments

A periodically poled (PP) structure was fabricated by a so-called field-cooling method with current control. The periodicity was set around 6.4 μm so as to generate 355 nm by THG of 1064 nm. The calculated periodicity based on the reported Sellmeier's equation is 6.39 μm. The size of poled area was 13 mm in length and 5 mm in width on the c-plate. The PP-domain pattern was revealed by chemical etching, and it was found that the PP pattern with almost 50:50 duty ratio was successfully fabricated in a whole device area.

Device performances of the fabricated PP-LBGO were examined by measuring sum frequency (SFG) conversion excited collinearly by both the fundamental 1064nm (pulsed with a repetition rate of 10 kHz and pulse width of 7.5 ns) and SHG-532 nm from a PP-MgO:SLT (stoichiometric $LiTaO_3$) device. Figure 1 shows an experimental setup schematically. The devices are designed based on the nonlinear components of d_{33}(=0.57pm/V) of LBGO.

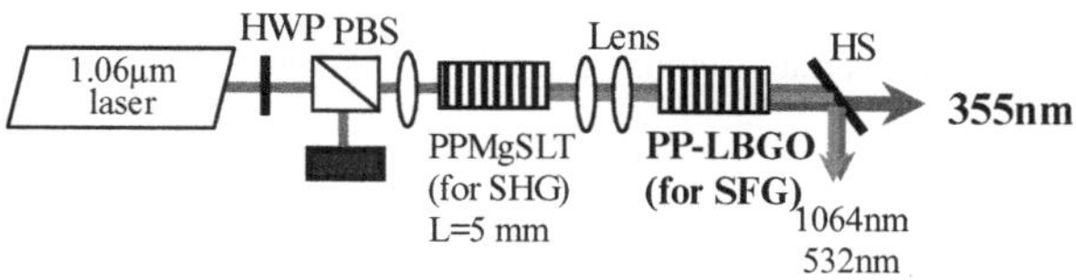

Figure 1. A schematic drawing of the experimental setup for THG.

RESULTS

1. Crystal growth and characterization

Prior to the crystal growth experiments, the degree of supercooling was considered. A large supercooling sometimes results in uncontrollable crystallization in Cz-pulling, because of a low thermal conductivity associated with a high viscosity of melt. The LBGO melt exhibited a large supercooling of ~50°C even when the cooling rate of the melt was as low as 0.1°C/min. A clear linear-relation between the degree of supercooling *ln*(ΔT) and the cooling rate *ln*(dT/dt), mentioned by a modified Nyvlt's semi-empirical equation[9], was confirmed for the first time for melts, as shown in Fig.2.

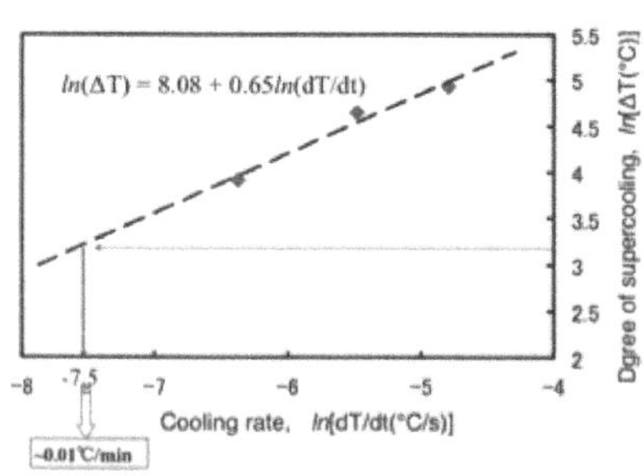

Figure 2. A linear relationship between the degree of supercooling ΔT and the cooling rate dT/dt of the LBGO melt (see the text)

It is well known that the growth velocity of crystallite depends on ΔT and there exists the maximum growth velocity normal to the crystal habit plane at the specific ΔT given by roughly

0.02×Tm (Tm; melting temperature), according to the classical heterogeneous nucleation experiment[10]. Since the melting temperature of LBGO is about 1200°C, the maximum growth velocity was calculated to be about 24°C of ΔT. We deduce the maximum cooling rate of about 0.01°C/min, based on the *ln*(ΔT) vs. *ln*(dT/dt) linear-relation shown in Fig.2. The appropriate normal growth rate is quite important to make larger the crystal diameter, as is described latter. In the present Cz-growth experiment, we carefully conducted temperature control of the melt during the crystal pulling with appropriate cooling rates less than the above rate so as to avoid poly-crystallization.

By optimizing the growth conditions, i.e. pulling/crystal rotation rates, atmospheres and cooling rates of the melt, we succeeded in growing the single crystals with a diameter as large as ~18mm in diameter and ~45mm in length, as shown in Fig.3(a). The as-grown crystals were wholly colorless, transparent without any macroscopic defects and non-hygroscopic. This non-hygroscopic property is very practical for wafer preparation processes. They exhibited a three-hold hexagonal shape on a cross-section, which reflects the crystal symmetry, as shown in Fig.3(b), and were surrounded by macroscopically smooth habit planes. When the cooling rates of the melt were not appropriate, poly-crystallization occurred frequently on these habit planes.

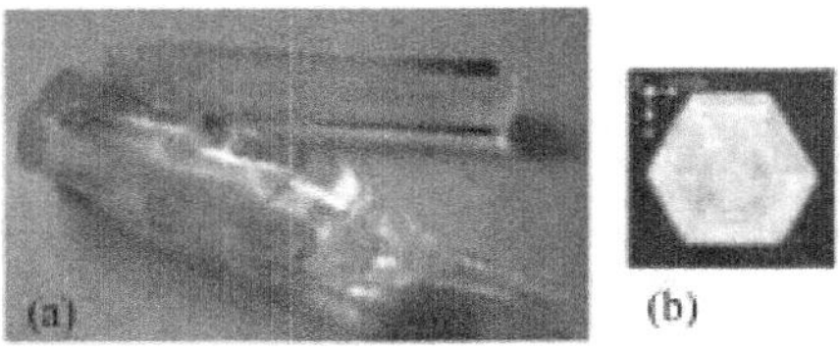

Figure 3. (a) typical as-grown LBGO single crystals, and (b) the (001) cross-section of the crystal.

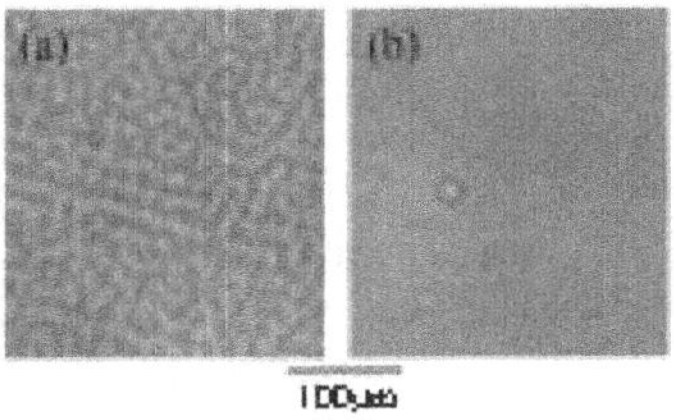

Figure 4. Ferroelectric domain structures of (a) as-grown and (b) poled wafers.

Figure 4(a) shows an example of ferroelectric multi-domain structure revealed on the c-plane of the as-grown crystal, where labyrince-like structure is observed. The single domain shown in Fig.4(b) was obtained by the field-cooling technique by applying the external dc field of more than 400 V/cm. We measured an optical transmittance from 190 nm to 3.3 μm with a 1.0 mm thick (001) c-plate. Although the optical transmittance was already reported1, the precise absorption edge was not described. Figure 5 shows the observed transmittance spectrum, and the inset shows the optical absorption edge lying at around 195nm. Among many ferroelectric crystals, LBGO has the shortest optical absorption edge, except some fluorides such as $BaMgF_4$[11]. This wide transparency is one of the most characteristic features of LBGO, and very suitable for generating UV wavelengths by the PP-QPM devices.

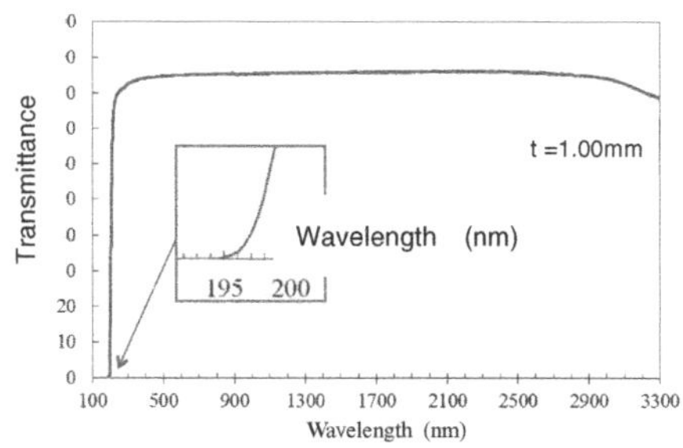

Figure 5. The optical transmittance of (001) c-plate.

Crystallinity was investigated by the X-ray rocking-curve, and by using the optical polarization microscope. Figure 6 shows the ω-scanned rocking-curve of (003) plane of the c-plate. The FWHM was as narrow as 12.2 arcsec. over the 15mmΦ wafer, which proves a very

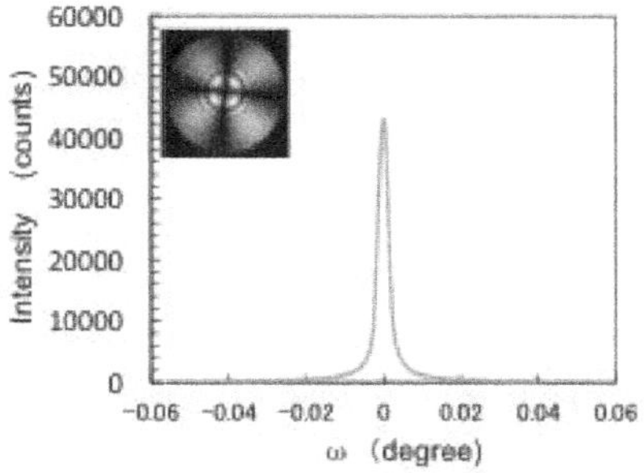

Figure 6. An X-ray ω-scanned rocking- curve of {003} reflection. The inset shows an optical conoscope figure.

high crystallinity. The inset in Fig.6. shows a typical conoscope figure, indicative of an optically uniaxial crystal. A clear isogyre without remarkable distortion is a proof of high optical quality. Crystal homogeneity was evaluated by measuring the Curie temperature Tc along the crystal length by differential scanning callorimetry (DSC) technique. The difference in Tc between the top and bottom parts of the crystal boule was confirmed to be only 0.1°C at

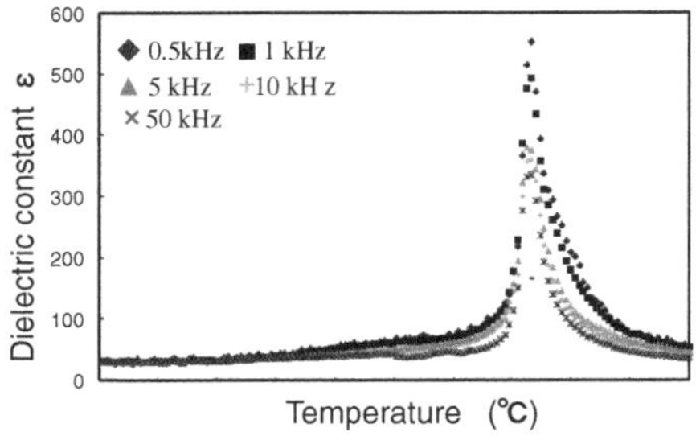

Figure 7. The temperature dependence of dielectric constant at different frequency

Tc~530°C, which indicates high compositional homogeneity over the crystal. Figure 7 demonstrates the temperature dependence of dielectric constant ε at different frequencies. Here, we notice that the peak temperature for each dependence does not vary remarkably around 530°C. This behavior suggests that LBGO is a stable second-order ferroelectric crystal.

2. Sum frequency generation (SFG) performances

As shown in Fig.1, both 1064nm fundamental beam and 532nm-SHG beam from a PP-MgSLT device were focused collinearly into the fabricated PP-QPM LBGO in order to generate a third harmonic 355nm generation, i.e., sum-frequency generation. The optical nonlinear effective coefficients deff is estimated to be about 0.36 pm/V, which is larger than those of $BaMgF_4$[11].

Device performances were verified by quadratic sum frequency generation of 1064nm, as shown in Fig.8. The output beam shape was almost circular because of non-walk-off, as shown in the inset. We confirmed more than 500mW output power of 355nm from this PP-LBGO. Although the conversion efficiency seems to be saturated, the conversion efficiency was achieved more than 10% from fundamental power and more than 20% as a SFG performance. The saturation of the conversion efficiency should be mainly caused by the multiplier effect associated with a saturation of SHG efficiency and a relatively low thermal conductivity of

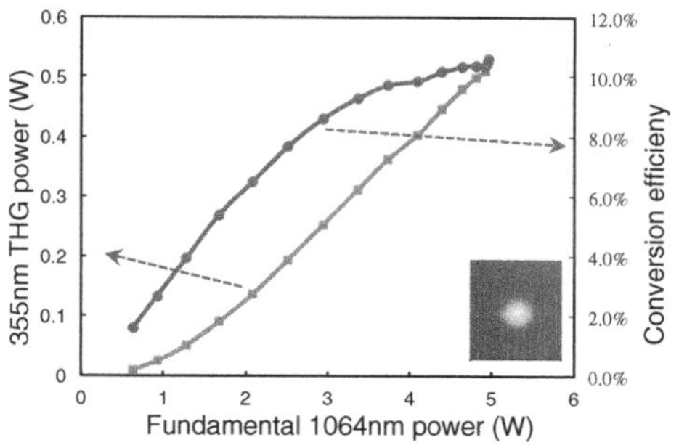

Fiure.8. Characteristics of 355nm generation and the conversion efficiency vs. the fundamental 1064nm input power. The inset shows an output beam shape.

LBGO which arises the so-called thermal lens, and further insight is now in progress.

From the results on 355nm generation experiments, transparent ferroelectric LBGO is strongly expected to be a good candidate for realizing 355 and also 266nm by PP-QPM structures without a walk-off.

SUMMARY

For fabricating the periodically poled, quasi phase- matching (PP-QPM) devices of ferroelectric $LaBGeO_5$ (LBGO), the Czochralski-pulling growth of the crystals adequate to devices was investigated. Based on the brief considerations about a large supercooling of the melt, we established the growth conditions suitable for growing *c*-axis-oriented, relatively large LBGO single crystals with high optical quality. The typical grown crystals were more than 18mm in diameter and more than 45mm in length. Crystallinity was confirmed to be12.2 arcsec. of the FWHM of ω-scanned X-ray rocking curve, and crystal homogeneity was evaluated by measuring a Curie temperature difference along the crystal length to be 0.1°C. The optical absorption edge was verified to be around 195nm.

The PP-QPM devices were fabricated successively, and about 500mW output power of 355nm generation was confirmed by mixing 1064nm and its second harmonic 532nm beams through the quadratic sun frequency conversion process. The conversion efficiency was more than 10% from fundamental power and more that 20% as a SFG performance.

Transparent and non-hygroscopic ferroelectric $LaBGeO_5$ is expected to be a good candidate for realizing 355 nm generation without a walk-off by PP-QPM structures.

ACKNOWLEDGMENTS

The authors express their thanks to to Prof. T.Taniuchi, Tohoku University, for his helpful collaboration of THG experiments, and also to Prof. Y.Uesu and Dr. H.Yokota, Waseda University, for supporting dielectric measurements.

REFERENCES

[1] J.Capmany, D.Jaqu, J.Garcia Sole and A.A.Kaminski, "Continuous wave laser radiation at 524nm from a self-frequency-doubled laser of $LaBGeO_5:Nd^{3+}$", Appl.Phys.Lett., vol.72, pp.531-533 (1998)

[2] *for example*; Y.Uesu, N.Horiuchi, E.Osakabe, S.Omori and B.A.Strukov, "On the phase transitions of new ferroelectric $LaBGeO_5$", J.Phys.Soc.Japan, vol.62, pp.2522-2523 (1993)

[3] J.Hirohashi, M.Hatori, M.Sakairi, S.Miyazawa, S.Takekawa, T.Taniuchi and Y.Furukawaet, "Second harmonic UV generation by novel periodically poled ferroelectric", Advanced Solid-State Lasers Conference 2013 (Paris), AM3A2

[4] J.Hirohashi, M.Hatori, M.Sakairi, S.Miyazawa, S.Takekawa, T.Taniuchi and Y.Furukawa, "355 nm generation by fab-out PP-LBGO devices", Conference on Laser and ElectroOptics 2014 (San Jose), SM4I.7

[5] A.A.Kaminskii, A.V.Butashin, I.A.Maslyanizin, B.V.Mill, V.S.Mironov, S.P.Rozov, S.E.Sarkisov and V.D.Shigorin, "Pure and Nd^{3+}-, Pr^{3+}-ion doped trigonal acentric $LaBGeO_5$ single crystals", phy.sta.sol.(a), vol.125, pp.671-696 (1991)

[6] S.Stefanovich, B.Mill and V.Sigaev, "Processing and Characxterization of Ferroelectrics in the Stillwelllite Family", Ferroelectrics, vol.201, pp.285-294 (1997)

[7] S.Miyazawa, S.Ichikawa, Y.H.Liu, S.YJin, T.Mtauoka and H.Nakae, "A novel substrate $LaBGeO_5$ lattice-matched to InN", phys.stat.sol.(a) vol.208, pp1195-1198 (2011)

[8] M.Sakairi, J.Hirohashi, S.Takekawa, S.Miyazawa and Y. Furukawa, "Czochralski growth of ferroelectric $LaBGeO_5$ single crystals for UV-generation", 6th Int'l Workshop on Cryst.Growth Technol., 2014 (Berlin), P2.10

[9] K.Sangwal, "A novel self-consistent Nyvlt-like equation for metastable zone width determined by the polythermal metod", Cryst.Res.Technol., vol.44, pp.231-47 (2008)

[10] G.J.Davies, Solidification and Casting, Appl.Sci.Publ., London, 1973, pp.25

[11] E.G.Villora, K.Shimamura, K.Sumiya and H.Ishibashi, "Birefringence- and quasi phase-matching with $BaMgF_4$ for vacuum-UV/UV and mod-IR all solid-state lasers", Opt.Exp., vol.17, pp.12362-12378 (2009)

THE GROWTH OF POTASSIUM TANTALATE NIOBATE ($KTa_xNb_{1-x}O_3$) SINGLE CRYSTAL BY VERTICAL BRIDGMAN METHOD

Toshinori Taishi[1,2], Kazuya Hosokawa[2], Keigo Hoshikawa[2], Takahiro Kojima[3], Junya Osada[3], Masahiro Sasaura[3], Yasunori Furukawa[3], Takayuki Komatsu[4]

[1] Center for Energy and Environmental Science, Shinshu University, 4-17-1, Wakasato, Nagano 380-8553, Japan

[2] Faculty of Engineering, Shinshu University, 4-17-1, Wakasato, Nagano 380-8553, Japan

[3] Oxide Corporation, 1747-1, Makihara, Mukawacho, Hokuto 408-0302, Japan

[4] NTT Advanced Technology Corporation, 3-1, Morinosato-wakamiya, Atsugi 243-0124, Japan

ABSTRACT

Growth of potassium tantalate niobate ($KTa_xNb_{1-x}O_3$) single crystals by the vertical Bridgman (VB) method was investigated. KTN crystals were grown using a $KTaO_3$ seed and K-rich raw materials in a platinum crucible in air. When raw material with an atomic ratio of Ta to Nb of 65:35 was used, a KTN crystal could be obtained in which the proportions of Ta and Nb varied widely. Colorless transparent cubic KTN crystallized above the seeding interface, where the atomic ratio of Ta to Nb was 89:11. It was found that this colorless transparent region was a single crystal. The color of the crystal changed to milky white and finally to powder-like white, and small cracks were observed following the milky white region. The proportions of Ta and Nb in the grown crystal gradually changed with the progress of growth, and the terminal of the crystal was tetragonal KTN. The tendency of the composition to change qualitatively obeys the binary phase diagram of $KTaO_3$-$KNbO_3$.

INTRODUCTION

Potassium tantalate niobate ($KTa_xNb_{1-x}O_3$: KTN) has a perovskite structure and is a complete solid solution of $KTaO_3$ and $KNbO_3$[1,2]. KTN (x<0.95) has three phase transition points, from cubic to tetragonal, from tetragonal to orthorhombic, and from orthorhombic to rhombohedral, whose temperatures change with the ratio of Ta to Nb in the KTN crystal[2,3]. A Ta-rich KTN crystal is a cubic structure, colorless and transparent, and $KTaO_3$ (x = 1) has no phase transition points. It is well known that KTN crystal displays a very large quadratic electro-optic (EO) effect (Kerr effect), and a KTN EO scanner using the space-charge-controlled mode of electrical conduction has been proposed[4,5]. Various EO applications using KTN with interesting optical properties will surely be proposed in the future.

However, the growth of large KTN single crystals is difficult. Typical crystal growth experiments with KTN, including growth methods and growth conditions, were summarized by Rytz et al.[2]. Almost all growth experiments were carried out from a K-rich or K_2CO_3-rich melt with a K content between 50 and 70 mol%, and the top-seeded solution growth (TSSG) method has typically been adopted. The results indicate that KTN is an incongruent material. Recently, a large rectangular parallelepiped KTN crystal about 30 mm long was grown that was colorless, transparent, and surrounded by {100} facet planes[6].

Crystal growth by the vertical Bridgman (VB) method has some advantages from the viewpoint of the production of large ingots with high reproducibility and without intentional diameter control. Recently, growth of langasite ($La_3Ga_5SiO_{14}$) single crystals[7] and sapphire (Al_2O_3) [8] by the VB method were reported. In our previous study, growth of $KNbO_3$ [9] and $KTaO_3$ [10] by the VB method was reported. When the K_2CO_3 content in the raw materials was at 52 mol%, $KNbO_3$ crystals could be grown without any production of $K_4Nb_6O_{17}$, which crystallizes as the primary phase from a stoichiometric $KNbO_3$ melt (with a K_2CO_3 content of 50 mol%). In the case of $KTaO_3$, when the K_2CO_3 content of the raw materials was 53 mol%, $KTaO_3$ crystals could be grown without $K_6Ta_{10.8}O_{30}$ appearing, which crystallizes as the primary phase from a stoichiometric $KTaO_3$ melt (with a K_2CO_3 content of 50 mol%). With regard to KTN crystals, growth by the VB method has not been reported.

This paper reports KTN crystal growth by the VB method using a $KTaO_3$ seed and K-rich raw materials in air. The ratio of Ta to Nb in the raw materials used was 65:35 in the present study. The crystallinity of crystals grown is evaluated. The compositional profile in crystals grown is also evaluated, and the results are discussed in terms of the binary phase diagram of $KTaO_3$-$KNbO_3$.

EXPERIMENTAL

Powders of K_2CO_3, Ta_2O_5 and Nb_2O_5 were used as starting materials, and were mixed in ethanol with the ratio of K_2CO_3 to (Ta_2O_5+Nb_2O_5) equal to 53:47 and with the Ta_2O_5 to Nb_2O_5 ratio 65:35. Dried mixed powders were calcined and then compressed to produce cylindrical raw materials. A [001]-oriented $KTaO_3$ single crystal seed, grown by the VB method[10], was set in a cylindrical platinum crucible (inner diameter 20 mm), and raw materials were placed on the seed inside the crucible. A platinum lid was put on the crucible, and a PtRh thermocouple was set at the bottom of the crucible to achieve suitable seeding temperature. An illustration of the arrangement around the crucible is shown in Fig. 1. Such a crucible was set on an alumina support in a furnace. The crystal growth was carried out in air. After the raw materials melted completely without melting of the seed, the crucible was pulled down at a rate of 1 mm/h. The rotation speed of the crucible was 5 rpm. After the completion of crystal growth, the crystal was cooled to room temperature for 16 h. The crystal grown was taken out of the crucible by peeling off the crucible. The growth apparatus and detailed growth conditions have been reported previously[7,10].

The as-grown crystals were evaluated by powder X-ray diffraction (XRD) using a RINT2200V/PC (RIGAKU) using CuK_α radiation. Wafers 0.3 mm in thickness were cut vertically parallel to (100) from the grown crystals and were polished on both faces. The

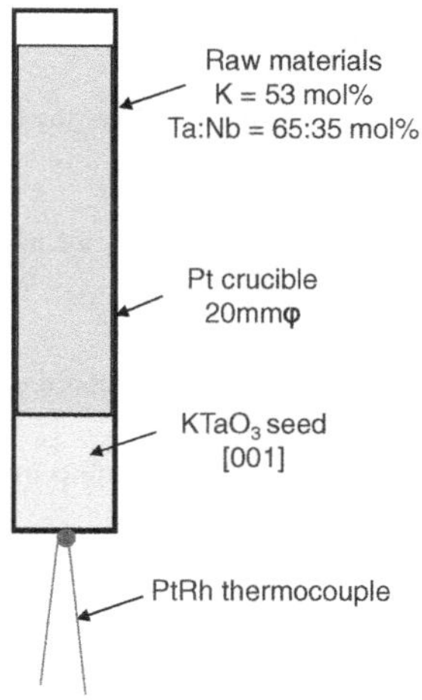

Fig. 1 The Pt crucible used in the present VB crystal growth.

composition of the grown crystals was evaluated by electron probe microanalysis (EPMA) using an EPMA-1610 (SHIMADZU). The composition of some specimens was also analyzed by inductively coupled plasma atomic emission spectroscopy (ICP) using an SFS1700 (Hitachi High-Tech Science). The quality of the grown crystal was evaluated using crossed polarizers, and some of the wafers were examined by X-ray topography using an XRT-300 (RIGAKU) with $g = 100$ diffraction using MoK_α radiation.

RESULTS AND DISCUSSION

Fig. 2 shows a photograph of a crystal grown using a $KTaO_3$ seed and K-rich raw materials with a Ta:Nb ratio of 65:35. An arrow indicates the seeding interface between the $KTaO_3$ seed (below) and the grown crystal (above). The length of both grown crystals was about 53 mm. The grown crystal just above the seeding interface, with a length of about 20 mm, is colorless and transparent. In this region, four flat planes were observed on the sides of the crystal, probably {100} faces. Then, the coloration of the crystal changed to milky white for a length of about 20 mm, and finally a white powder-like region was observed with a length of about 13 mm. This final region consisted of a hygroscopic white alkaline compound, which could easily be removed in water.

Specimens obtained from positions labeled as positions (i), (ii) and above position (iii) in Fig. 2 were analyzed by powder XRD, and results are shown in Fig. 3. XRD patterns of specimens from (i) and (ii) are similar to that of $KTaO_3$, indicating a cubic structure. On the other hand, twin peaks were detected in each peak of the sample from above (iii), and this was similar to $KNbO_3$. These results indicate that a KTN crystal can be grown by the VB

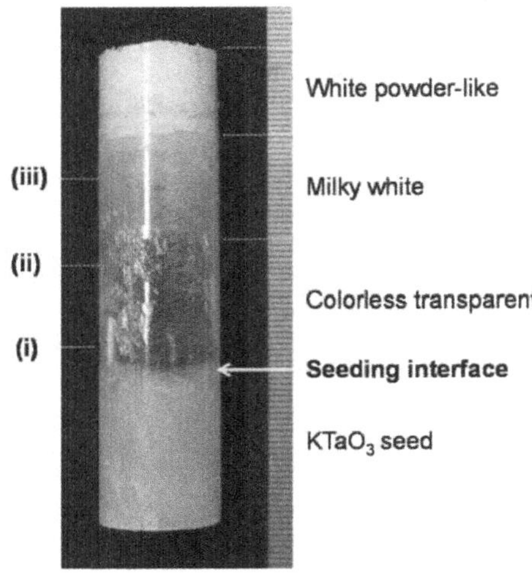

Fig. 2 Photograph of a KTN crystal grown using a [001] $KTaO_3$ seed and K-rich raw materials with the ratio of Ta to Nb = 65:35.

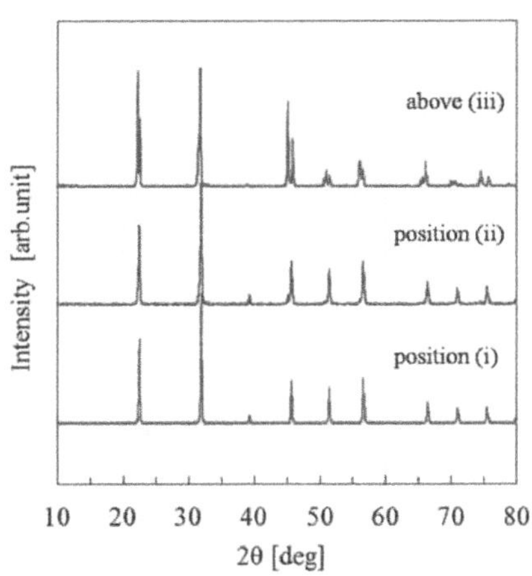

Fig. 3 XRD patterns for positions (i), (ii) and above the position (iii), labeled in Fig. 2, in a grown KTN crystal.

method but crystal structures in the grown KTN crystal differ at different positions in the crystal.

Fig. 4 (a) shows a photograph of a polished wafer cut vertically from the crystal shown in Fig. 2. In the colorless transparent region, few cracks were observed, but small cracks were observed in the milky white region. Such cracks increased, and finally the coloration became

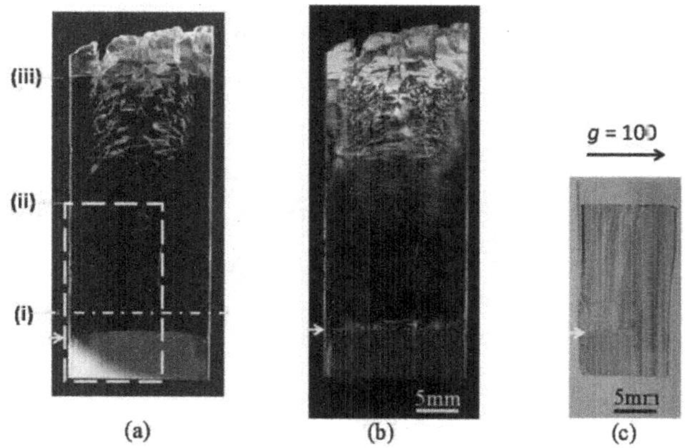

Fig. 4 (a) A photograph of the KTN wafer cut vertically along the growth direction, (b) a photograph of the wafer viewed between crossed polarizers, (c) an X-ray topographic image of the wafer portion shown in a dotted rectangle in Fig. 4 (a).

cloudy at the terminal end of the crystal. Fig. 4 (b) is a photograph of the wafer shown in Fig. 4 (a), as viewed between crossed polarizers. In the colorless transparent region above the seeding interface, no other phases and no grain boundary were observed, but the image became cloudy below the appearance of the small cracks in the milky white region, indicating that strain was induced near the boundary between the colorless transparent and milky white regions. The quality of grown crystal became poor as growth progressed. An X-ray topographic image of the region surrounded by the white dotted rectangle is shown in Fig. 4 (c). Although many vertical straight lines were observed in the topograph, a diffraction image of the whole of the specimen was obtained, indicating that this region is single KTN crystal. Many images of lines were probably related to the threading dislocation, which came from the $KTaO_3$ seed and were induced by the difference in lattice constant between the $KTaO_3$ seed and the grown crystal just above the seeding interface. Analyzed by ICP, the ratio of Ta to Nb just above the interface at the position labeled (i) in Fig. 2 was 89:11. This corresponded to about 0.03% of difference in lattice constant at the interface. To grow higher quality KTN crystal, the preparation of a higher quality $KTaO_3$ seed is important.

Fig. 5 shows the compositional profile of Ta and Nb in the wafer shown in Fig. 4 (a), as analyzed by EPMA. Below the seeding interface, only Ta was detected, while Nb was lower than the detection limit. Then, Nb at about 10 mol% was detected just above the interface, while the content of Ta rapidly decreased, indicating that the KTN crystal was grown just above the $KTaO_3$ seed. Then, the content of Ta and Nb along the growth direction gradually changed, with Ta decreasing while Nb increased. The gradient of this change at position (iii) was larger than that at positions (i) and (ii). Profiles crossed each other near position (iii), and the composition reached about 90 mol% Nb, while Ta approached 10 mol% at the terminal

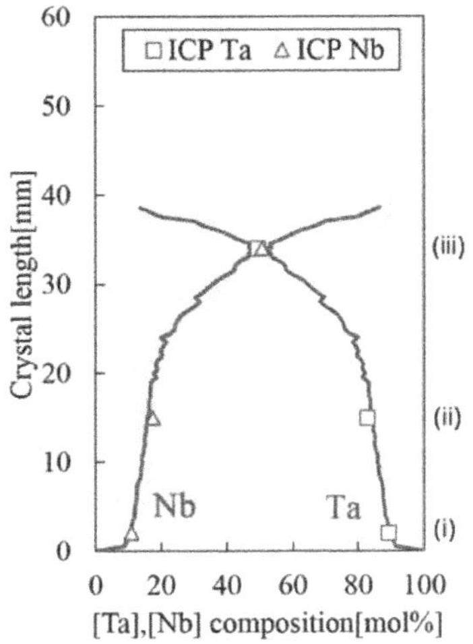

Fig. 5 Composition profiles of Ta and Nb in the KTN wafer shown in Fig. 4 (a) analyzed by EPMA. Open squares and open triangles represent compositions of Ta and Nb analyzed by ICP, respectively.

end of the crystal. ICP analysis was also carried out on three specimens cut from different positions in the crystal (positions (i), (ii) and (iii) in Fig. 2), and the results were plotted using open squares for Ta and open triangles for Nb. Compositions analyzed by EPMA and ICP were in good agreement with each other.

Thus, a KTN crystal could be grown by the VB method, notably including a colorless transparent region above the seeding interface that was a single crystal. In addition, such a KTN crystal has a unique composition profile for Ta and Nb, which gradually undergoes a wide change in the KTN crystal. Such a gradual composition change with respect to the proportions of Ta and Nb can be explained by the phase diagram of $KTaO_3$-$KNbO_3$. Experimental results with the reported binary phase diagram[2] are summarized in Fig. 6. The initial composition is shown as position (i), then as crystal growth progressed, the content of Ta and Nb changed from Ta-rich to Nb-rich along liquidus and solidus lines in the phase diagram, and finally the crystal became $KNbO_3$. Positions (i), (ii) and (iii) correspond to those shown in Fig. 2. The lattice constant of KTN increases with decreasing Ta or increasing Nb content. Assuming that the lattice constant changes linearly between $KTaO_3$ (a = 3.9941 Å) and orthorhombic $KNbO_3$ (a = 4.0160 Å)[2], the differences in lattice constant between position (i) and (ii) corresponds to 0.02% and between (ii) and (iii) corresponds to 0.18%. The strain and small cracks shown in Figs. 4 (a) and (b) are probably related to a larger lattice constant change following the milky white region in the crystal. In addition, at position (iii) a phase transition from cubic to tetragonal occurs. At room temperature (25°C), the atomic ratio of Ta to Nb in KTN for the phase transition from cubic to tetragonal is about 65:35, and the

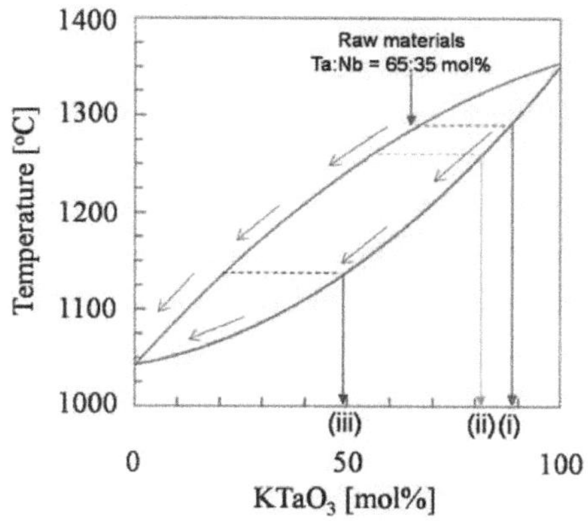

Fig. 6 Composition change of Ta and Nb in the KTN crystal in the present study. Data were plotted on the reported binary phase diagram of $KTaO_3$-$KNbO_3$ [2].

atomic ratio of Ta to Nb at position (iii) was 49:51, that is, this position is tetragonal KTN with lower quality than that in the colorless transparent cubic region.

One of the features of the VB method is that it grows a crystal from all melt in a crucible. In the present study, this produced a KTN crystal with a large and gradual composition change and with a phase transition point at room temperature. Such a crystal, with a monocrystal-like quality, is quite unusual. Currently, investigation is underway on the detailed composition changes of Ta and Nb in the crystal and the comparison of that variation with the binary phase diagram. The growth of KTN crystals using various raw materials with different relative quantities of Ta and Nb is also being conducted. In addition, several efforts to grow a VB-KTN crystal with homogeneous composition and no compositional change in the melt have been carried out, and will be reported elsewhere.

CONCLUSION

The growth of KTN crystals by the VB method was investigated. Using a $KTaO_3$ seed and K-rich raw materials with an atomic ratio of Ta to Nb of 65:35, KTN crystals were grown in a platinum crucible in air. A KTN crystal whose proportions of Ta and Nb varied widely from one region to another was obtained. Above the seeding interface, colorless transparent cubic KTN crystal was obtained with an atomic ratio of Ta to Nb of 89:11. X-ray topographic examination showed that the colorless transparent region was a single crystal. The coloration of the crystal changed to milky white and finally to powder-like white, and small cracks were observed following the milky white region. The proportions of Ta and Nb in the grown crystal gradually changed with increasing solidification, and the terminal end of the crystal was tetragonal KTN. The composition change was discussed with reference to the binary phase diagram of $KTaO_3$-$KNbO_3$, with which it was in qualitative agreement. VB-grown KTN crystal with monocrystal-like quality and a gradual composition change is very uncommon. If high quality long KTN crystals with a homogeneous composition can be obtained, VB growth will be very useful for EO applications.

ACKNOWLEDGEMENTS

This work was supported in part by an Adaptable and Seamless Technology Transfer Program through Targetdrive R&D (A-STEP: AS2315011C) from the Japan Science and Technology Agency.

REFERENCE

[1] Reisman A., Triebwasser S., Holzberg F., "Phase Diagram of the System $KNbO_3$-$KTaO_3$ by the Methods of Differential Thermal and Resistance Analysis", J. Am. Chem. Soc., 77, 4228-4230, 1955.

[2] Rytz D., Scheel H. J., "Crystal Growth of $KTa_{1-x}Nb_xO_3$ ($0<x\leq0.04$) Solid Solutions by a Slow-Cooling Method", J. Cryst. Growth, 59, 468-484, 1982.

[3] Triewasser S., "Study of Ferroelectric Transitions of Solid-Solution Single Crystals of $KNbO_3$-$KTaO_3$" Phys. Rev., 114,63-68 1959.

[4] Nakamura K., Miyazu J., Sasaura M., Fujiura K., "Wide-Angle, Low-Voltage Electro-Optic Beam Deflection Based on Space-Charge-Controlled Mode of Electrical Conduction in $KTa_{1-x}Nb_xO_3$ ", Appl. Phys. Lett., 89, 131115, 2006.

[5] Nakamura K., Miyazu J., Sasaki Y., Imai T., Sasaura M., Fujiura K., "Space-Charge-Controlled Electro-Optic Effect: Optical Beam Deflection by Electro-Optic Effect and Space-Charge-Controlled Electrical Conduction", J. Appl. Phys. 104, 013105, 2008.

[6] Imai T., Sasaura M., Nakamura K., Fujiura K., "Crystal Growth and Electro-optic Properties of $KTa_{1-x}Nb_xO_3$", NTT Tech. J., 5, 1-8, 2007.

[7] Taishi T., Hayashi T., Fukami T., Hoshikawa K., Yonenaga I., "Single-Crystal Growth of Langasite ($La_3Ga_5SiO_{14}$) by the Vertical Bridgman (VB) Method in Air and in an Ar Atmosphere", J. Cryst. Growth, 304, 4-6, 2007.

[8] Miyagawa C., Kobayashi T., Taishi T., Hoshikawa K., "Demonstration of Crack-Free *c*-axis Sapphire Crystal Growth Using the Vertical Bridgman Method", J. Cryst. Growth, 372, 95-99, 2013.

[9] Kudo K., Kakiuchi K., Endo N., Bamba N., Hoshikawa K., Fukami T., "Effect of Polind Field on Piezoelectric Properties of $KNbO_3$ Crystal Growth by Vertical Bridgman Method", Jpn. J. Appl. Phys., 43, 6701-6705, 2004.

[10] Taishi T., Hosokawa K., Hosokawa K., Kojima T., "The Growth of Potassium Tantalate ($KTaO_3$) Single Crystals by Vertical Bridgman Method", submitted to J. Cryst. Growth.

GROWTH OF $Y_3Al_5O_{12}$ SINGLE CRYSTALS VIA EDGE-DEFINED FILM-FED GROWTH TECHNIQUE USING MO CRUCIBLES

T. Tokairin[1], J. Hayashi[2], G. Villora[3], K. Shimamura[3]
[1]Ibaraki University, Hitachi, Japan
[2]Shinko Manufacturing Co., Ltd., Osaka, Japan
[3]National Institute for Materials Science, Tsukuba, Japan

ABSTRACT

With the principal objective of providing inexpensive Ce:YAG single-crystal phosphors for higher brightness white LEDs, we grew Ce:YAG single crystals using the edge-defined film-fed growth (EFG) technique with a Mo crucible. From the analysis on the adhesive material and blackish crystals, the crystals have a blackish color by carbon evaporation. Transparent and crack-free Ce:YAG crystals were obtained by control of carbon evaporation. The relations among optical absorption, crystal defects and growth conditions were examined and discussed. Furthermore, examined the relationship between the temperature gradient, growth speed, and the segregation coefficient. The Ce: YAG single crystals produced in this manner have potential application in higher brightness white LEDs.

INTRODUCTION

High-power light-emitting diodes (LEDs) first appeared in 1994[1], with white LEDs being developed later, in 1998[2]. Current commercial white LEDs are based on blue LEDs with added yellow phosphor powder; this consists of $Y_3Al_5O_{12}$ (YAG) powder doped with a cerium compound. In recent years, many studies on higher brightness white LEDs have been conducted. These LEDs are designed for use in projectors and car headlights, which require higher luminance than conventional devices. However, high-power LEDs and laser diodes (LDs) require thermally stable phosphor, and a recent study has reported that single-crystal Ce:$Y_3Al_5O_{12}$ (Ce:YAG) plates constitute phosphors with the required thermal stability[3]. Ce:YAG single crystals are conventionally grown using the Czochralski (CZ) technique, in which Ir metal is used as the crucible material. However, this requirement is restricting the development of cost-effective Ce:YAG single-crystal growth methods because of the expensiveness of Ir crucible. Furthermore, high-speed growth is difficult, because a segregation coefficient of Ce is small in the CZ technique. In 1971[4], the edge-defined film-fed growth (EFG) method, which has a large segregation coefficient, was proposed. The EFG method was later evaluated through a study using sapphire, and it was found that this process is capable of producing crystals with almost any cross-sectional shape or size[5]. At present, Mo metals are being widely used in the growth processes of sapphire single-crystals via either EFG or Kyropoulos techniques, although the application of Mo to the CZ growth of YAG has not been very successful to date[5,6]. In this work, the application of Mo to YAG growth using the EFG technique is attempted. Then, the factors leading to the poor performance of this technique are identified and countermeasures to improve the performance are taken. Finally, the crystals obtained with these countermeasures are evaluated, and the applicability of this approach to production is also examined.

EXPERIMENTAL

Figure 1 (a) shows the crystal growth furnace used in the present investigation. This carbon resistance-heating furnace has a hot zone containing a Mo reflector and carbon insulator. Figure 1 (b) shows a simulation of the apparatus, produced using the PHOENIX thermal field analysis software.

(a) (b)

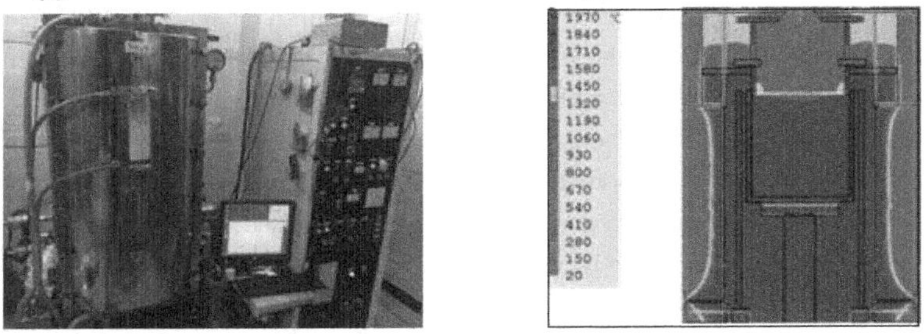

Figure 1. Equipment and Simulation. (a) is Resistance-heating crystal-growth equipment. (b) is Simulated apparatus using PHOENIX analysis software.

A Mo crucible with 50 mm external diameter, 1.5 mm thickness, 50 mm height, and 47 mm depth was used, along with Mo dies with 40 mm width, 30 mm height, and 1.5 mm thickness, which contained 0.5 mm wide slits. A Mo lid and cylinder were used to minimize the heat radiation from the heat source. The starting material was high-purity α-Al_2O_3, Y_2O_3, and CeO_2 powder (99.99%), and the Ce concentration in the melt was varied from 0 to 2 at%. The charge weight of the YAG raw materials was 30 g, and a pull-up speed V for the crystal growth was 3–100 mm/h. The weight of grown crystal was 28-g crystals, and the distance from the melt surface to the top of the die was 25 mm. The target temperature gradient at the tops of the dies was determined to be 10–40 °C/cm without crystal growth; this was achieved by adjusting the crucible position. The temperature was measured using a B-type thermocouple. The growth rate was varied between 3 and 100 mm/h. The crystals, which were polished to 0.6–0.8 mm thickness, were annealed and their absorption spectra were obtained using a JASCO Model V-50 spectrophotometer. The Ce concentrations in the Ce:YAG crystals were determined using a corrected calibration curve obtained via glow discharge mass spectroscopy (GDMS) analysis and the 340 nm absorption spectra of the crystals.

ANALYSIS AND DISCUSSION

Crystals were initially grown at a speed of 8 mm/h in an Ar atmosphere; however, the first obtained crystal was blackish. The surface of the blackish Ce:YAG crystal was analyzed using a HITACHI TM3030 EDX and the results are shown in figure 2. It can be seen that the ratio of Al to Y atoms is 0.48 in the top region and 1.26 at the tail. Figure 2-(d) and (e) shows the results of the analysis of the adhesion substance on the graphite heater and seed. Yttrium was only found at the carbon heater, whereas aluminum carbide was found at the seed.

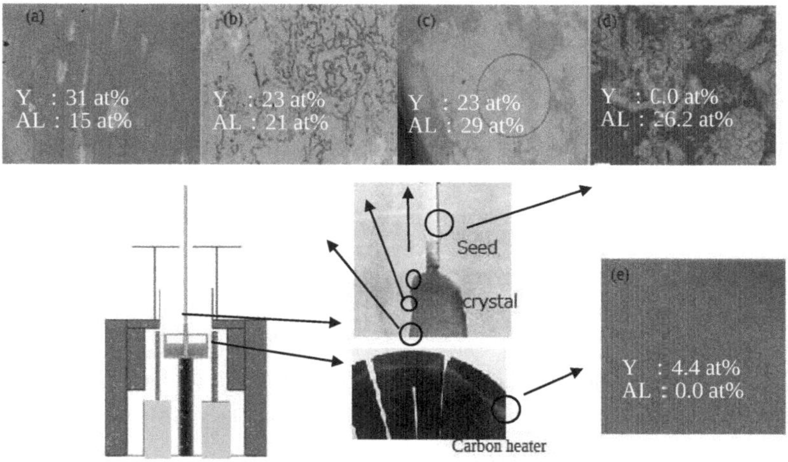

Figure2. Elemental analysis on the crystal and seed, heater using EDX.
(a) is tail of crystal,(b) is middle, (c) is top, (d) is aluminum carbide at the seed, (e) is yttrium carbide at the heater.

These results indicate that the raw materials and carbon reacted chemically and evaporated in the initial growth process. Figure 3 shows the absorption spectrum of the blackish YAG crystal as measured by the spectrophotometer. It is apparent that the absorption in the UV wavelength range was stronger than that of YAG crystals grown via the CZ technique using Ir crucibles, and this was not improved by annealing at 235nm. The blackish appearance of the crystal was due to carbon evaporation, which is increased by the presence of water and oxygen in the growth atmosphere. For countermeasures of the problem, we promptly evacuated the water and oxygen in the chamber using a vacuum pump and then replaced with Ar gas. Both room temperature and 1000°C operations were performed, and the crucible was covered with a Mo plate so that the melt was not degraded by carbon evaporation. The obtained crystals exhibited high absorption of short wavelengths; therefore, they were heat annealed at 1300 °C for 2 h. Figure 4 shows the UV transmission spectra of the YAG crystals grown using the anti-carbon-evaporation measures. It is apparent that increased transparency was achieved using this approach. Further, it can be seen that YAG crystals with quality comparable to that of YAG single crystals obtained via the CZ technique using Ir crucibles were produced.

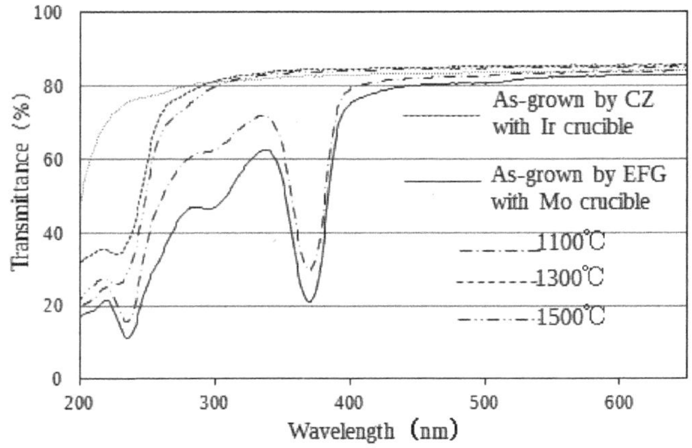

Figure 3. UV transmission spectrum of YAG crystals grown by CZ technique using Ir crucible , compared to grown by EFG technique using a Mo crucible. The crystals was annealed in air at temperatures of 1300 °C or above.

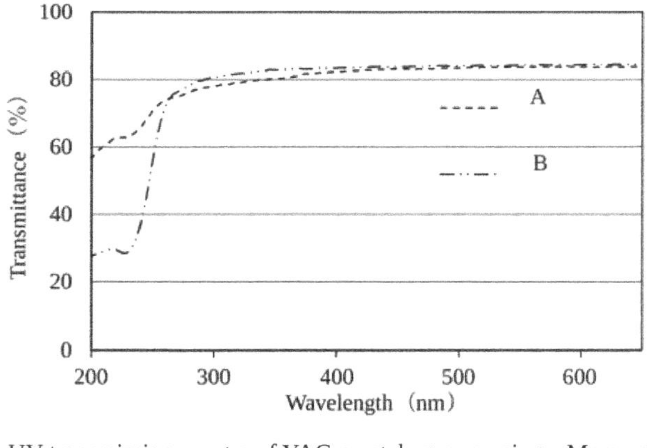

Figure 4. UV transmission spectra of YAG crystals grown using a Mo crucible. Transmittance curve (A) was achieved by a reduction of carbon evaporation, while curve (B) shows decreased transmission because of carbon evaporation.

It was also confirmed that transparent and crack-free Ce:YAG single crystals were obtained, as shown in Figure 5. Thus, Ce:YAG single crystals suitable for phosphor applications with 25-mm width and 150-mm length were grown using Mo dies and crucibles.

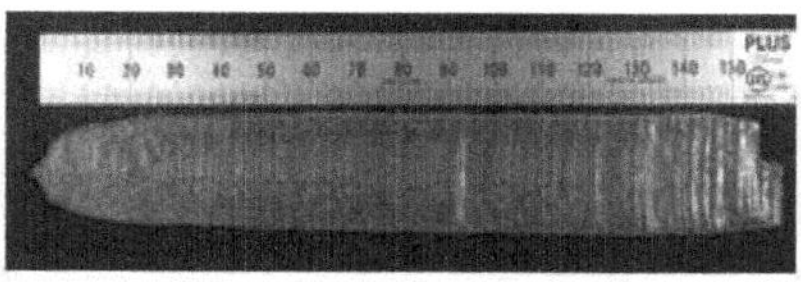

Figure 5. Ce:YAG single crystals grown by the EFG technique using Mo crucibles.

Table 1
Impurities (in mass ppm) in the crystal and materia

Element	Starting material	Crystal
Mo	ND	15
Si	10	0.42
Fe	5	0.5

The segregation coefficient k for the EFG method is expressed in Eq. (1). The higher the growth rate and the greater the increase in the temperature gradient, the greater the increase in the segregation. The crystal concentration equations for the EFG and CZ methods are

FG method : $C_s = C_0\{(1-k)[1-\exp(-k \times V \times g \times D)]+k\}$ (1)

CZ method : $C_s = C_0(1-g)^{k-1}$ (2)

Where V is the growth rate, D is the diffusion factor, C_s is the crystal concentration, g is the solidification fraction (0–1), k is the effective segregation coefficient, and C_0 is the initial melt density. On this occasion, an attempt was made to decrease the diffusion coefficient as much as possible by setting the slit width to 0.5 mm or less, and by fixing the distance from the liquid surface to the top of the die to 30 mm.

Figure 6 illustrates the variation of the segregation coefficient k of the Ce:YAG single crystals when the temperature gradient and V were varied. Although a higher segregation coefficient was obtained compared to the CZ method (Eq. (2)) with k = 0.1–0.2, a high concentration gradient through the crystal cross-section was observed. Figure 9 shows the Ce concentration gradients for Ce:YAG crystals grown using the EFG and CZ methods, for various solidification fraction(g).

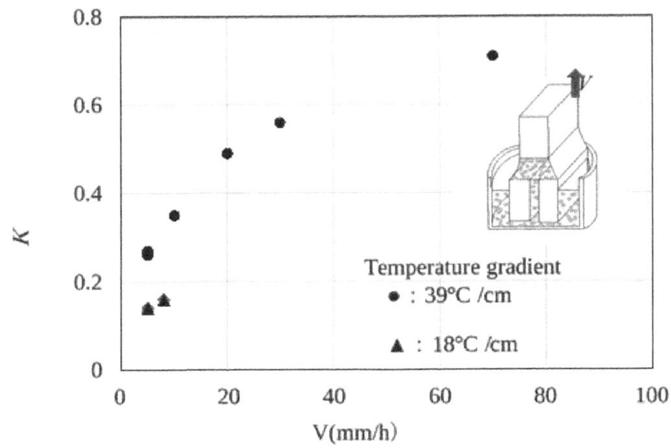

Figure 6. Segregation coefficient of Ce:YAG by EFG method for various pull-up rates. The k of crystals grown at a temperature gradient of 39°C/cm is greater than that at 18°C/cm.

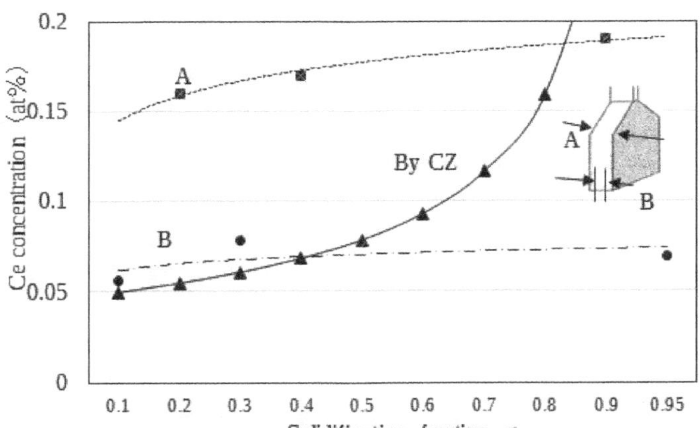

Figure 7. Ce concentration gradient in Ce:YAG grown via EFG and CZ methods.
Ce concentration in the central part (B) of crystal is smaller than whole(A).

As regards defects, etch pitting was observed for 30 min etching with phosphoric acid heated to 250 °C, and a micrograph of an etch pit is shown in Figure 8. It was discovered that point defects form when the growth rate increases, and micropipes from air bubbles occur when the temperature is further increased. High Ce concentration can therefore occur near the crystal substrate surface as a result of Ce becoming trapped in the point defects.

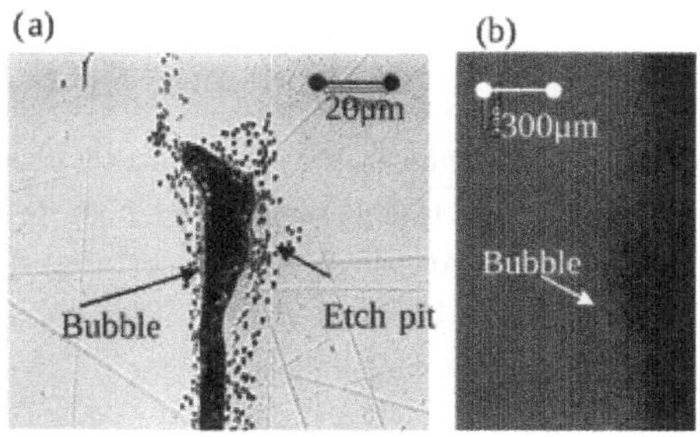

Figure 8. Optical photomicrographs of crystal defects in Ce:YAG crystals. (a) is Etch pits by etching with phosphoric acid. (b) is Bubbles made on the crystal surface.

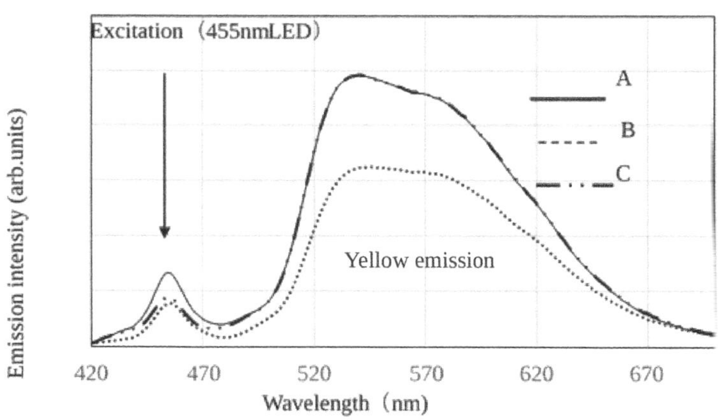

Figure 9. Photoluminescence of Ce:YAG crystals grown using Ir and Mo crucibles. Curve (A) is for as-grown crystals from an Ir crucible, (B) is for as-grown crystals from a Mo crucible, (C) is annealed crystal of (B).

Figure 9 provides a quantum efficiency comparison for crystals grown using Ir and Mo crucibles. Although the internal quantum efficiency is almost identical in both cases, there is a decrease in the amount of light transmitted by the Mo-grown crystal for the 455 nm excitation source, because a slightly higher amount of absorption occurs on the short-wavelength side. The almost identical quantum efficiency is because of an increase in the amount of fluorescence.

CONCLUSIONS

Transparent and crack-free Ce:YAG crystals were obtained via the EFG technique using Mo crucibles, enabling suppression of carbon evaporation. The segregation coefficient was larger than that obtained for crystals produced using the CZ technique. Therefore, high-speed growth was enabled and the production of economical Ce:YAG single crystals for use as single-crystal phosphors was realized. We believe that the Ce:YAG crystals obtained using this technique have useful application as phosphors for higher brightness white LEDs.

ACKNOWLEDGEMENTS

The authors would like to thank the National Institute for Materials Science (NIMS) and Shinko Manufacturing Co., Ltd., for their support.

REFERENCES

[1] Nakamura, S., Mukai, T., & Senoh, M. (1994). Candela-Class High-Brightness InGaN/AlGaN Double-Heterostructure Blue-Light-Emitting Diodes. *Applied Physics Letters, Volume 64,* 1687–1689.

[2] Bando, K., Sakano, K., Noguchi, Y., & Shimizu, Y. (1998). Development of High-Bright and Pure-White LED Lamps. *Journal of Light and Visual Environment, Volume 22,* 2–5.

[3] Arjoca, S., Villora, G., Inomata, D., Aoki, K., Sugahara, Y., & Shimamura, K. (2014). Ce:$(Y_{1-x}Lu_x)_3Al_5O_{12}$ Single-Crystal Phosphor Plates for High-Brightness White LEDs/LDs with High-Color Rendering (Ra > 90) and Temperature Stability. *Materials Research Express, Volume 1,* 025041.

[4] LaBelle, Jr., H. E. (1980). EFG, the Invention and Application to Sapphire Growth. *Journal of Crystal Growth, Volume 50,* 8–17.

[5] Kvapil, Ji., Kvapil, Jo., Manek B., Perner, B., Autrata, R., & Schauer, P. (1981). Czochralski Growth of YAG:Ce in a Reducing Protective Atmosphere. *Journal of Crystal Growth, Volume 52,* 542–545.

[6] Kvapil, Ji. Kvapil, Jo., Manek, B., Blazkk, K., & Hendrichi, Z. (1985). Nonstoichiometric Defects in YAG and YAP. *Crystal Research and Technology, Volume 20,* 473–478.

Multifunctional Coatings for Energy and Environmental Applications

NANOPARTICLE PASTE INJECTION INTO GAS-FLAME THERMAL SPRAY FOR SPEEDY CERAMIC COATING

Soshu Kirihara
Joining and Welding Research Institute, Osaka University,
11-1 Mihogaoka Ibaraki, 567-0047 Osaka, Japan

ABSTRACT

Thermal nanoparticle spraying was newly developed to create fine ceramic layers with a high deposition rate. Nanoparticles were dispersed into liquid resins at approximately half volume fraction, and the obtained high viscosity slurry was blown as micromist through an acetylene gas-flame spray. Microstructures were observed using scanning electron microscopy. Further, X-ray diffraction spectroscopy was used to analyze the residual carbon elements produced by imperfect combustions of the liquid resin. Effective dielectric constants of the ceramic layers were measured successfully using a terahertz time domain spectroscopic device to calculate the porosities as the volume content of air defects. In addition, the microstructural modulations of the coated layers through the spraying parameters controls are discussed.

INTRODUCTION

Suspension plasma spraying techniques have been actively investigated to effectively introduce nanometer-sized particles dispersed with liquid solvents into plasma or gas flames[1,2]. Fine coated layers can be created successfully on practical alloy substrates using dilute suspensions of 5-10 % volume content nanoparticles[3,4]. However, it is difficult to increase the deposition rates of the conventional thermal spraying processes using material powders that are the size of micrometers[5,6]. In our investigation, thermal nanoparticle spraying was newly developed to create fine ceramic layers without microcracks and pores on the free surfaces of the practical alloy substrates[7,8]. Resin slurries with ceramic nanoparticles were used as raw materials for ceramic coating. The paste materials show thixotropic fluid properties, and therefore, their viscosities can be controlled by the introduction of shear forces. The nanoparticle slurries were sputtered using a compressed air jet, and the micromist droplets were blown into arc plasma flows with an argon gas sprayed from the vertical direction. The introduced nanoparticles were heated, and they accelerated effectively through the center core of the plasma flows. However, nanoparticles introduced into the side flows of the plasma flame were insufficient to increase the thermal and kinetic energies. Recently, slurry micromists were introduced into high temperature gas flames in a coaxial direction using a Rokide spraying gun as shown in Fig. 1. In this investigation, the fluid viscosities of the mixt pastes were measured and optimized to realize homogenized particle dispersions and smooth slurry flows. The relationship between the slurry properties and the formation of the microcracks and pores in the coated layers will be discussed.

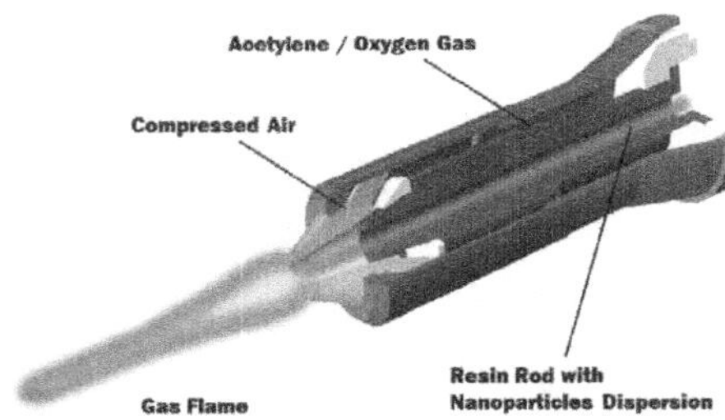

Fig. 1 Schematic illustration of solid rod feeding process in Rokide flame gun system. Nanometer-sized ceramic particles dispersed in resin matrix with large volume fraction can be introduced into high temperature flame coaxially to realize a high deposition rate.

EXPERIMENTAL PROCEDURE

Spherical-shaped alumina particles (TM-5D, Taimei Chemicals, Japan) with an average diameter of 200 nm were dispersed into the acrylic liquid resins (KC1287, JSR, Japan) at 40 vol%. These obtained slurries were mixed in airtight containers through the planetary configuration of rotation and revolution movements (SK-350T, Shashin Kagaku, Japan) as shown in Fig. 2. The nanoparticle-dispersed resin paste with a volume of 200 ml and five alumina balls with diameters of 5 mm were packed into a cylindrical container. Degassing and mixing were executed by revolution and rotation at speeds of 700 and 300 rpm, respectively. The mixing times were changed at 300, 600, and 900 s. The fluid characteristics of the slurry paste were evaluated using a viscosity and viscoelasticity measuring instrument (VT550, Thermo Scientific, USA). The paste material was inserted between two metal disks as shown in Fig. 3. Shear stresses were loaded from the rotating upper disk to the paste on the lower disk and were measured with a torque meter. The variations of the shear stress according to the shear velocity were measured dynamically. The thixotropic slurry was injected perpendicularly into a gas flow chamber through a syringe nozzle with a diameter of 4 mm, and the fluid viscosity decreased owing to shear stress loading as schematically illustrated in Fig. 4. The slurry mists were sprayed on a glass plate to measure the droplet diameters and observe the spattering patterns. Micromists were formed using the compressed air jet with a gas pressure of 2 atm and were introduced into the acetylene flame with a gas pressure of 0.15 MPa as shown in Fig. 1. The alumina-coated layer was formed on a SUS-304 stainless steel substrate with dimensions of 50 × 50 × 5 mm, placed at a distance of 150 mm from the flame gun. The crystal phase of the coated layer was analyzed using X-ray diffraction, while its cross-sectional microstructures and compositional distributions were observed using optical and scanning electron microscopy. Effective dielectric constants of the ceramic layers were measured successfully using a terahertz time domain spectroscopic device to estimate the porosities as the volume contents of nano-sized air defects.

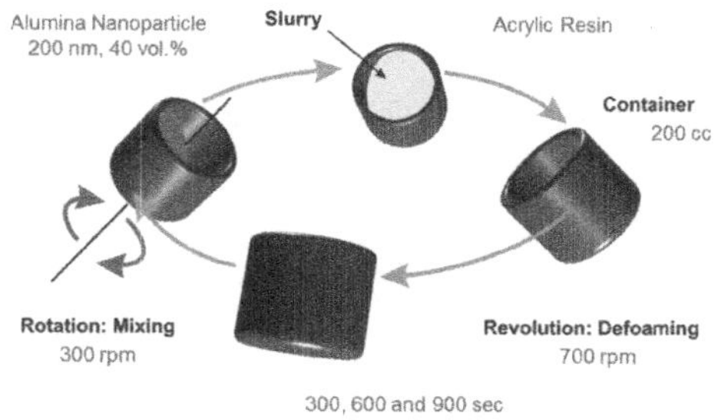

Fig. 2 Schematic illustration of a planetary mixing machine. Revolution and rotation movements of the packed container realize degassing and particle dispersions, respectively.

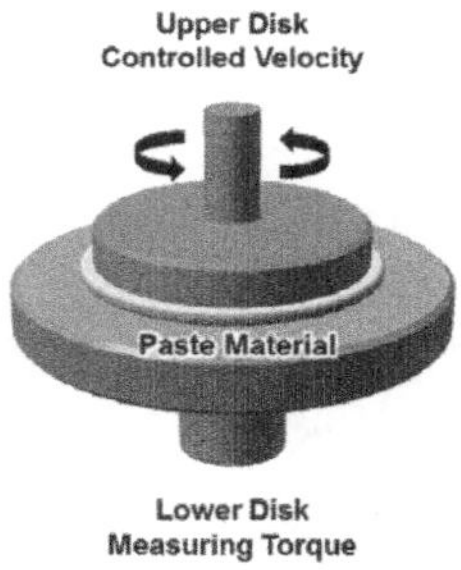

Fig. 3 Schematic illustration of a viscosity and viscoelasticity measuring. Shear stresses are loaded from the rotating upper disk on the lower one and are measured with a torque meter.

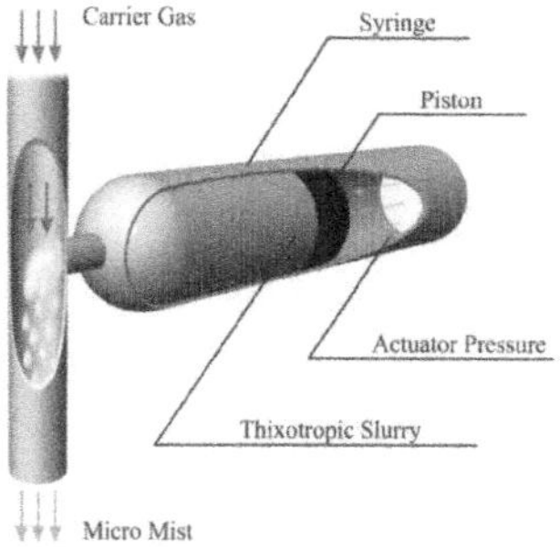

Fig. 4 Schematic illustration of a slurry mist supplying. The thixotropic slurry including nanoparticles was injected into a gas flow chamber perpendicularly to spatter microdroplets.

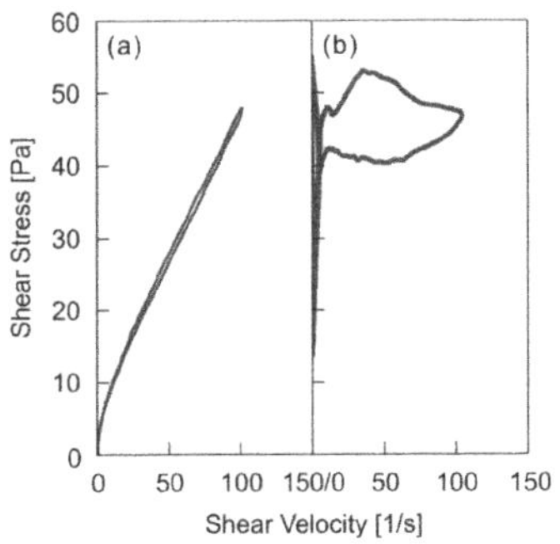

Fig. 5 Relationship between the shear velocity and stress of alumina nanoparticles dispersed in acrylic resins at mixing times of (a) 300 s and (b) 900 s.

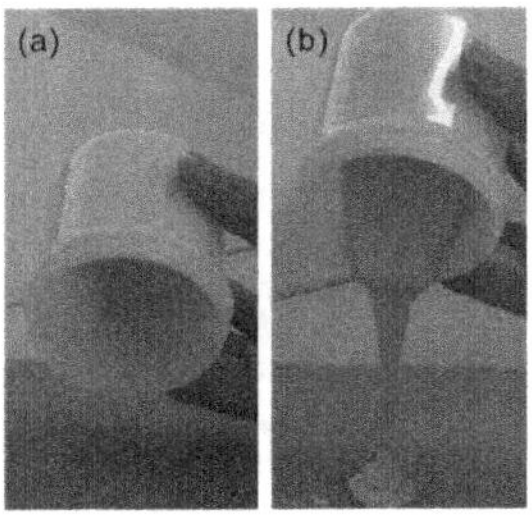

Fig. 6 Thixotropic rheological profiles of non-Newtonian nanoparticle-dispersed resins as mixed (a) and after shear stress introduction (b).

RESULTS AND DISCUSSION

The dynamic variations in the shear stress and velocities of the nanoparticles paste were measured as shown in Fig. 6. For a shorter mixing time (300 s), the shear stress increased linearly according to the share velocity as shown in Fig. 5 (a); thus, the typical Newtonian fluid property was observed. The particles were considered to coagulate in the resin solvent of the Newtonian liquid[9]. In contrast, for a longer mixing time (900 s), the paste material exhibited different rheological profiles at the increasing and decreasing share velocities as shown in Fig. 5 (b). The resin paste exhibited high viscosity as shown in Fig. 6 (a) by the nanoparticle surface fictions, though smooth fluid flows were observed as shown in Fig. 6 (b) by loading the shear stress to disband the particle interactions. The non-Newtonian fluid properties are especially called thixotropic characteristics, and these phenomena indicate the homogeneous dispersion of nanoparticles in the resin solvent[10]. The formed slurry with the particle dispersion of 42 vol%

could not flow smoothly. For supplying stable materials and for continuous mist creation, the maximum volume content of the ceramic nanoparticles can be optimized to 40 %. In this slurry fluid, it is theoretically considered that nanoparticles disperse homogeneously and move spatially without coagglutination. The obtained slurry was sprayed on the glass plate to observe and measure the shape and size of the mist droplets. These droplet diameters became finer as shown in Fig. 7, and the variations reduced by decreasing the slurry supply rate at 130 cc/min. The minimum diameter size was 50 μm. Figure 8 shows the fine alumina-coated layer formed on a stainless steel substrate. The ceramic layers with a thickness of 50 μm were formed at a supply rate of 300 gpm. Structural defects such as cracks or pores were not noticeable through microscopic observations. The coated ceramic layer and alloy substrates were joined successfully without the defects of voids or exfoliations in the interfaces. The X-ray diffraction analysis did not detect contaminations of the carbon element in the created alumina layers derived from the acrylic liquid resins. The dielectric constants of the coated layers were measured by electromagnetic waves reflection spectroscopy. The measured value was lower than the ideal dielectric constant of sintered alumina. The coated layer was considered to include nanometer-sized air cavities.

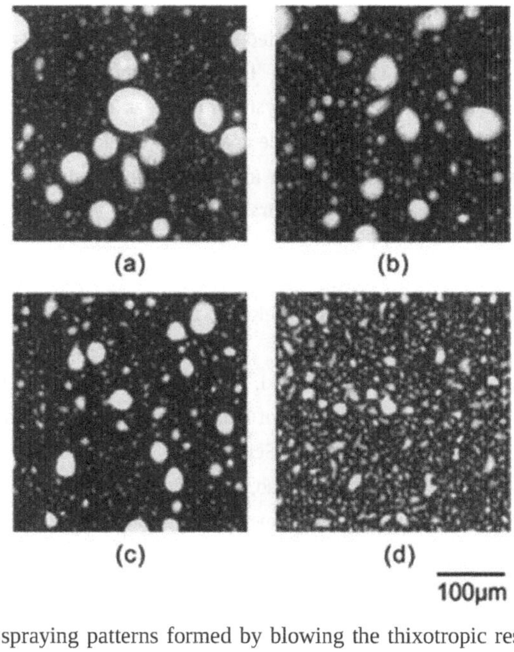

Fig. 7 Micromist spraying patterns formed by blowing the thixotropic resin slurry containing alumina nanoparticles.

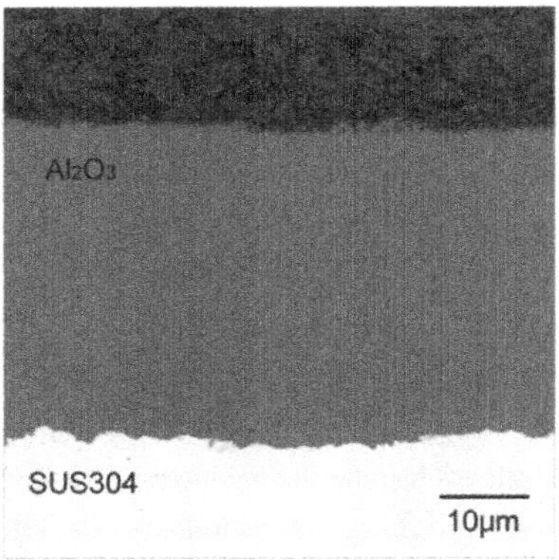

Fig. 8 Cross-sectional microstructure of the alumina-coated layer formed on a stainless steel substrate by nanoparticle thermal spraying. The relative density of a coated layer estimated by a terahertz time domain spectroscopic device was lower than the sintered alumina bulk.

CONCLUSIONS

Fine alumina-coated layers were created successfully on stainless steel substrates using thermal nanoparticle spraying. Using these techniques, thixotropic slurries including the nanometer-sized particles were formulated systematically as raw materials. The slurry paste modulations are key techniques for nanoparticle coatings to create useful and functional surfaces on various components. In the near future, these investigated coating techniques will be efficiently used to realize effective coating without structural defects.

REFERENCES

[1]A. Killinger, M. Kuhn, R. Gadow, "High-Velocity Suspension Flame Spraying", Surf. Coat. Technol., 2006, 201, p 1922-1929.

[2]T. Boeltken, D. Soysal, S. Lee, G. Straczewski, U. Gerhards, P. Peifer, J. Arnold, R. Dittmeyer, "Perspectives of Suspension Plasma Spraying of Palladium Nanoparticles for Preparation of Thin Palladium Composite Membranes" J. Membr. Sci., 2014, 468, p 233-241.

[3]P. He, H. Sun, Y. Gui, F. Lapostolle, H. Liao, C. Coddet, "Microstructure and Properties of Nanostructured YSZ Coating Prepared by Suspension Plasma Spraying at Low Pressure", Surf. Coat. Technol., 2015, 261, p 318-326.

[4]P. Carpio, E. Rayón, L. Pawłowski, A. Cattini, R. Benavente, E. Bannier, M. Salvador, E. Sánchez, "Microstructure and Indentation Mechanical Properties of YSZ Nanostructured Coatings Obtained by Suspension Plasma Spraying, Surf. Coat. Technol., 2013, 220, p 237-243.

[5]R. Rampon, O. Marchand, C. Filiatre, G. Bertrand, "Influence of Suspension Characteristics on

Coatings Microstructure Obtained by Suspension Plasma Spraying", Surf. Coat. Technol., 2008, 202, p 4337-4342.

[6]H. Kaβner, R. Vaβen, D. Stöver, "Study on Instant Ddroplet and Particle Stages During Suspension Plasma Spraying (SPS)", Surf. Coat. Technol., 2008, 202, p 4355-4361.

[7]S. Kirihara, "Development of Thermal Nanoparticles Spraying Technique", Therm. Spray Technol., 2010, 30, p 44-50.

[8]S. Kirihara, "Introduction of Ceramics Nanoparticles into Thermal Spraying", J. Jap. Weld. Soc., 2010, 80, p 6-9.

[9]A. Beris, E. Stiakakis, D. Vlassopoulos, "A thermodynamically Consistent Model for the Thixotropic Behavior of Concentrated Star Polymer Suspensions", J. Non-Newtcnian Fluid Mech., 2008, 152, p 76-85.

[10]C. H. Lee, V. Moturi, Y. Lee, "Thixotropic Property in Pharmaceutical Formulations", J. Contr. Rel., 2009, 136, p 88-98.

CONTRIBUTION TO ELECTROCHEMICAL OXIDATION OF A XANTHENE DYE ONTO CU_2O THIN FILM ELECTRODE

M. El hajji, A. Tara, Ph. Dony, O. Jbara, L. Bazzi, A. Benlhachemi and N. Kireche Materials Science and engineering laboratory, University of Reims Champagne-Ardenne, Reims, France
Materials and environmental laboratory, Ibn Zohr University, Agadir, Morocco
IATOS of GRESPI team, UFR Exact and Natural Sciences, Moulin de la Housse campus, Reims, France

ABSTRACT

A facile synthesis route of cuprous oxide (Cu_2O) was introduced. Properties of obtained thin film varied as functions of reaction temperature. Morphology and structure of Cu_2O are analyzed by SEM, XRD and FTIR. The catalytic performances and reaction kinetics of Cu_2O catalyst for the degradation of Rhodamine B (RhB) are investigated. The results show pure Cu_2O electrocatalyst with triangular morphology is obtained and demonstrates good potential on degradation of RhB. The degree of RhB degradation is improved significantly in the presence of 0.1 M NaCl. The degradation rate constant decreases as the initial RhB concentration increases and the degradation reaction can be represented by pseudo-first-order kinetics model. These synthesized Cu_2O thin films would be highly attractive for practical applications in water pollutant removal and environmental remediation.

INTRODUCTION

The evolution of the color industry has been closely related to the discovery of mauveine (aniline, basic dye). Today there are more than 10.000 colors; this large number requires the establishment of a classification system. An index of English dyes was developed by the society of dyer and colorists and by the American Association of Textile Chemists and Colorists, which describes the essential characteristics of each dye such as: chemical formula, color, heat resistance, resistance to solvents, and a description of the main areas of use[1]. These organic dyes, coming from textile, cosmetic, paper, leather, pharmaceutical and nutrition industries, posed a significant threat to the environment and public health owing to their potential to form toxic aromatic products with carcinogenic and mutagenic properties[2-5]. Rhodamine B (RhB) is the most commonly employed basic dye. This actual dye-stuff is certainly carcinogenic and has been removed from cosmetic and food use. It is also most dangerous when ingested; also it can cause increased heart rate, shock, cyanosis, and tissue necrosis in humans. Various methods have been developed for the treatment of colored wastewater[6]. Among them, the advanced oxidation processes (AOP) have received much attention on the removal of various types of dye wastewater because of the generation of oxidizing radical species such as $^{\cdot}OH$ in the presence of catalysts, which can degrade a broad range of organic pollutants quickly and non-selectively[7]. Recently, many semiconductor metal oxides, such as TiO_2, ZnO, CeO_2 and Cu_2O, have been adopted as catalyst to degrade dye wastewater[8,9]. Copper oxide (I) is an important p-type semiconductor with a direct band gap 2.0-

2.2 eV, which is suitable for many applications, such as photocatalysis[10], gas sensors[11,12], and electrochemical cells[13,14]. In recent years, many efforts have been made to obtain Cu_2O nanomaterials[15], and several Cu_2O nanostructures have already been synthesized, including nanowires, nanospheres, nanocubes, nanomultipods, nanoboxes and nanocages[16,17]. Solvothermal synthesis[18-20], chemical reduction[21-24], electrodeposition[25-27], and irradiation[28,29]methods have been applied to prepare Cu_2O nanocrystals. In this work, copper oxide (I) thin films with uniform diameter can be obtained through electrodeposition method. Morphologies of Cu_2O prepared under different temperatures are displayed. In addition, electrocatalytic activities of Cu_2O are measured regarding the effect of various experimental parameters and operating conditions. The removal of the organics from the effluent was followed by chemical oxygen demand (COD).

EXPERIMENTAL

Synthesis of Cu_2O thin films

In a typical synthesis, 0.05 mol/L $Cu(NO_3)_2.3H_2O$ (Sigma Aldrich) was added into 100 mL of bidistilled water containing 0.1 M of tartaric acid under continuous stirring. The pH solution was adjusted by addition of a controlled amount of 1M KOH. A Titanium disk (Ti), Platinum sheet (Pt) and an Hg/Hg_2Cl_2 (satd. KCl) (SCE) were used as working electrode, counter electrode and reference electrode respectively. Films were grown for 60 minutes. The deposition bath temperature was controlled between 30 °C and 50 °C by a precision water bath. Fourier transform infrared spectra (FTIR) were recorded at room temperature using a Perkin Elmer 1600 series FTIR spectrophotometer. The XRD spectrums of the samples were recorded at room temperature, and the surface morphology of the films was analyzed using a scanning electron microscope. The same electrochemical equipments and all characterization details had been described elsewhere[30].

Rhodamine B degradation

The cyclic voltammetry (CV) analysis and the electrochemical measurements were performed using a computer controlled by Potentiostat/Galvanostat PGP 201 Radiometer Copenhagen Voltalab.TM associated to "Volta-Master" software. A conventional three electrodes cell (70 cm^3) thermoregulated glass cell was used. The anode was a disk plate of « Ti/Cu_2O » with an effective surface area of 1.2 cm^2, whereas the cathode was a platinum electrode. The distance between the anode and the cathode was set at 0.9 cm^2. A saturated calomel electrode (SCE) was used as a reference. The cell was connected to a water bath in order to control and keep the temperature constant; also, it was set on a magnetic stirrer for continuous mixing. The operating current density during degradation of RhB was 70 $mAcm^{-2}$. The degradation took 300 to 360 minutes. The chemical oxygen demand (COD) permitted the evaluation of the kinetic of organic matter rottenness. LOVIBOND Checkit COD VARIO device was used for all COD measurements.

RESULTS AND DISCUSSION

Principle and basic requirements for cathodic electrodeposition of Cu_2O

Generally, the simplified overall reaction of copper oxide formation by cathodic electrodeposition can be summarized as:

$$P(M)_{solution} + P(O)_{solution} + e^-_{substrate} \rightarrow MO_{substrate} + \text{soluble products} \quad (1)$$

With P(M) and P(O) being dissolved precursor species for the metallic element M and oxygen O in a solution. For Cu_2O, the mechanism of cathodic electrodeposition of oxide can be accompanied by the following steps.

• Nitrate reduction

$$NO_3^- + H_2O + 2e^- \rightarrow NO_2^- + 2OH^- \quad (2)$$

• O_2 reduction

$$O_2 + 2H_2O + 4e^- \rightarrow 4OH^- \quad (3)$$

• Copper reduction

$$Cu^{2+} + e^- \rightarrow Cu^+ \quad (4)$$

• H_3O^+ consumption and OH^- production favor the formation of Cu(OH)

$$Cu^+ + OH^- \rightarrow Cu(OH) \quad (5)$$

• Deposition of cuprous oxide

$$2\,Cu(OH) \rightarrow Cu_2O + H_2O \quad (6)$$

Morphology and structure of Cu_2O

Uniform Cu_2O with rough surface were obtained by the simple cathodic deposition method. Figure 1 shows SEM images of Cu_2O spheres prepared under different temperatures.

Figure 1. SEM images of Cu_2O obtained at 60°C (a), 40°C (b), and 20°C (c) at 3000x magnification.

It can be clearly observed that the sizes and morphologies of Cu_2O changed under different conditions. When the reaction takes place in hot water bath keeping at 60°C, the obtained Cu_2O are well distributed in size and they have an angular shape (triangle) with a small particle size (Figure 1a). The surface of the deposit is homogeneous, dense and uniform. Moderately lowering the bath temperature, the morphology changes completely. The surface of the deposit becomes less homogeneous and more, we find grain as truncated octahedron with a mean particle size (Figure 1b). At 20 ° C, we see an almost different morphology of the foregoing. The surface is composed of large grains deformed and grouped as agglomerates (Figure 1c). The change of the grain size in the deposit with the Cu_2O bath temperature is also confirmed in the results of the SEM.

The composition and phase purity of the products prepared at different reaction temperatures were characterized by XRD as shown in Figure 2. XRD patterns are indexed to Cu_2O (JCPDS 05-0667)[31], which show the expected, (111), (200), (220), (311), and (222) diffraction peaks corresponding to crystal planes of the Cu_2O crystals. Thus, no phase attributed to copper oxide (II) "CuO", metallic copper "Cu" or copper hydroxide appears for all three samples.

Fourier transforms infrared transmission spectra of Cu_2O are showed in Figure 3A. For all three samples, a main peak at about 620 cm^{-1} is depicted. The slight shift in peak position could be the effect of reduction in particle size. No infrared active mode corresponding to CuO could be seen. The sample was also free of OH type species. The energy dispersive analysis shows the near stoichiometric composition for films deposited onto various substrates (Figure 3B).

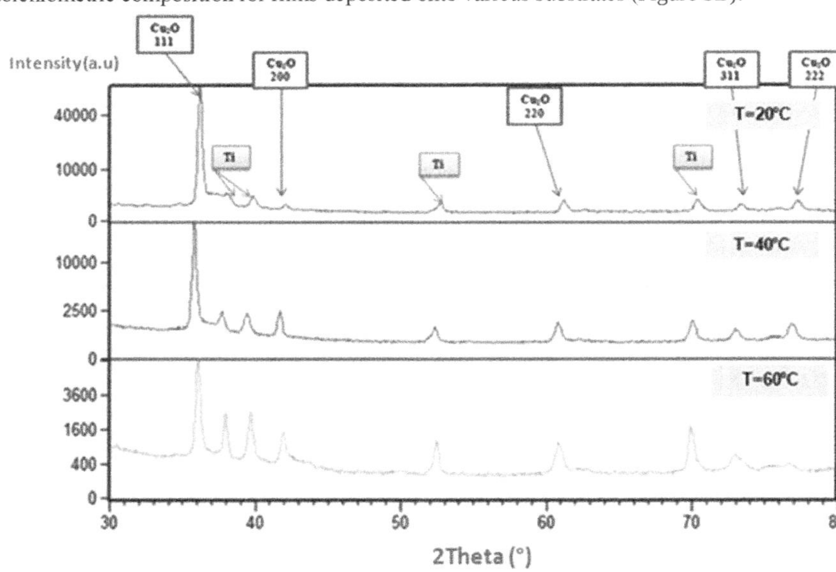

Figure 2. XRD patterns of Cu_2O prepared at 60°C, 20°C, and 40°C.

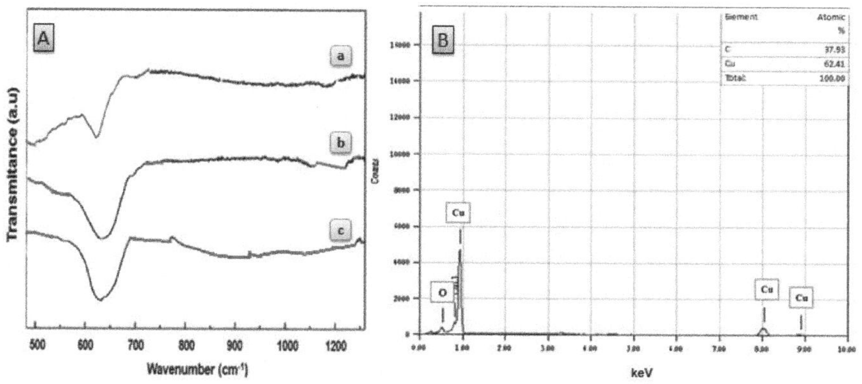

Figure 3. (A) Fourier transform infrared transmission spectra for Cu_2O prepared at (a) 60°C, (c) 20°C, and (b) 40°C nanocrystals showing the main Cu_2O vibrational mode at 620 cm^{-1}. (B) EDS spectrum related to the deposit of the pure Cu_2O (prepared at 60°C).

Parameters effect on degradation efficiency of Rhodamine B

Effect of NaCl concentration

To test the catalytic activities of obtained Cu_2O, RhB as cationic charged molecule was used in the electrodegradation experiments. In this part, the three kinds of Cu_2O thin films elaborated at different temperatures are tested.

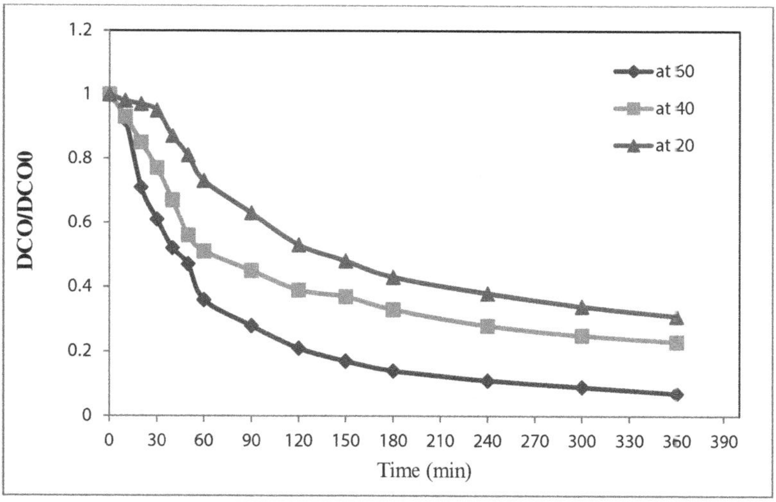

Figure 4. Effect of temperature growth (°C) of Cu_2O thin films used as electrode on degree of RhB degradation.

It is found that oxidation onto electrode based on Cu_2O thin film sample obtained at hot temperature (60°C) gives good results. Figure 4 shows the electrocatalytic performance of Cu_2O for the degradation of MO. The experimental results disclose that Cu_2O thin film obtained at hot temperature (60°C) allow superior catalytic activity. The influence of NaCl concentrations on the degree of RhB degradation is shown in Figure 5.

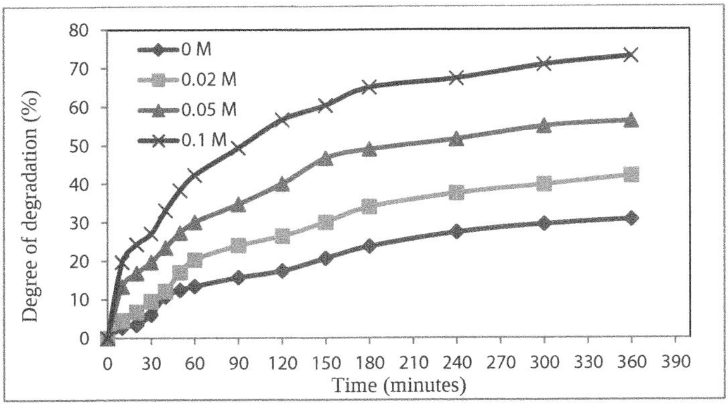

Figure 5. Effect of NaCl concentrations on degree of RhB degradation

It can be seen that the degree of RhB degradation increases sharply in the initial stage and then gradually reaches equilibrium. With the increase of NaCl concentration from 0.02 M to 0.1 M, the degree of RhB degradation increases from 30.7% to 73.03%.

Effect of RhB concentration

The influence of MB concentrations on the degree of MB degradation is shown in Figure 6.

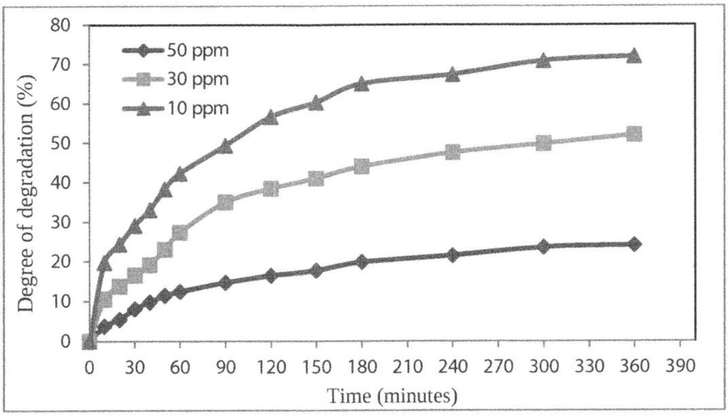

Figure 6. Effect initial RhB concentrations on degree of RhB degradation

It can be seen that the degree of RhB degradation increases sharply in the initial stage and then gradually reaches equilibrium after 240 min of degradation. With the increase of RhB concentration from 10 ppm to 50 ppm, the degree of RhB degradation decreases from 73.03% to 24.1%. Figure 7 shows the plots of Ln (DCO_0/DCO) versus time for RhB degradation at different concentrations ranging from 10 ppm to 50 ppm, which indicates that the degradation process follows the pseudo-first-order kinetics model.

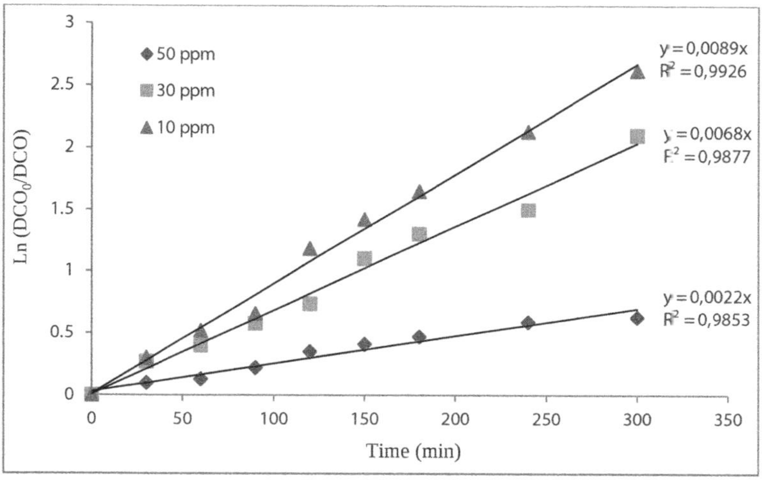

Figure 7. Pseudo-first-order kinetics model of RhB degradation by Cu_2O thin film electrode

The degradation rate constants are found to be in the following order: $8.9.10^{-3}$ min^{-1} (10 ppm) > $6.8.10^{-3}$ min^{-1} (30 ppm) > $2.2.10^{-3}$ min^{-1} (50 ppm), which is inversely in proportion to the RhB concentration (as shown in Table I).

Table I. Kinetics parameters of RhB degradation by Cu_2O thin film electrode

RhB concentration (ppm)	The degradation rate constants (min^{-1})	R^2
10	0.0089	0.992
30	0.0068	0.987
50	0.0022	0.985

The results may be attributed to that a relative lower concentration of ·OH produced in reaction system results in the decrease of the degree of RhB degradation when higher concentration of RhB is treated[32].

CONCLUSION

In summary, Cu_2O thin films are prepared successfully by a facile cathodic deposition method. The investigation on catalytic activity of as-prepared Cu_2O films demonstrates high degradation degree for RhB can be obtained using Cu_2O electrodeposited at high temperature (60°C) in the presence of NaCl (0.1 M). However, insufficiency of NaCl does not favor the further improvement of degradation degree of the pollutant. The degradation degree of RhB is dependent on its concentration. With the increase of RhB concentration from 10 ppm to 50 ppm, the degradation degree of RhB decreases from 73.03% to 24.1%. The degradation reactions follow the pseudo-first-order kinetics model. The elaborated Cu_2O thin films show in this work a good efficiency and still exhibit great competition to other electrodes. Future efforts would be directed to the development of electrode used as anode for the degradation of organic pollutants, in which doped Cu_2O is deposited on different substrates.

REFERENCES

[1] Perrin, R., pierre, S.J. (1999). Chimie industrielle. 2eme édition.

[2] Kalsoom, U., Ashraf, S. S., Meetani, M. A., Rauf, M. A., and Bhatti, H. N. (2012). Degradation and kinetics of H_2O_2 assisted photochemical oxidation of Remazol Turquoise Blue, *Chemical Engineering Journal, Volume 200*, 373-379.

[3] Bizani, E., Fytianos, K., Poulios, I., Tsiridis, V. (2006). Photocatalytic decolorization and degradation of dye solutions and wastewaters in the presence of titanium dioxide, *J. Hazardous Materials, volume 136*, 85–94.

[4] Mahmoodi, N.M., Arami, M., Limaee, N. Y. (2006). Photocatalytic degradation of triazinic ringcontaining azo dye (Reactive Red 198) by using immobilized TiO_2 photoreactor: Bench scale study, *J. Hazardous Materials, Volume 133,* 113–118.

[5] Crepy, M.N. (2004). Dermatoses professionnelles aux colorants, Fiche d'allergologie-dermatologie professionnelle hôpital Cochin.

[6] Hwang, K., Lee, J., Shim, W., Jang, H.D., Lee, S. and Yoo, S. (2012). Adsorption and photocatalysis of nanocrystalline TiO_2 particles prepared by sol–gel method for methylene blue degradation, *Advanced Powder Technology, Volume 23,* 414-418.

[7] Khataee, A.R., Vatanpour, V. and Amani Ghadim, A.R. (2009). Decolorization of C.I. Acid Blue solution by UV/Nano-TiO_2, Fenton, Fenton-like, electro-Fenton and electrocoagulation processes: A comparative study, *Journal of Hazardous Materials, Volume 161,* 1225-1233.

[8] Fink, L., Dror, I. and Berkowitz, B. (2012). Enrofloxacin oxidative degradation facilitated by metal oxide nanoparticles, *Chemosphere, Volume 86,* 144-149.

[9] Park, J., Jang, I., Kwon, B., Jang, S. and Oh, S. (2013). Formation of manganese oxide shells on silica spheres with various crystal structures using surfactants for the degradation of methylene blue dye, *Materials Research Bulletin, Volume 48,* 469-475.

[10] Xu, H.L., Wang, W.Z., Zhu, W. (2006). Shape evolution and size-controllable synthesis of Cu_2O octahedra and their morphology-dependent photocatalytic properties. *J Phys Chem, Volume 110,* 13829–13834.
[11] Zhang, J.T., Liu, J.F., Peng, Q., Wang, X., Li, Y.D. (2006). Nearly monodisperse Cu_2O and CuO nanospheres: preparation and applications for sensitive gas sensors, *Chem Mater, Volume 18,* 867–871.
[12] Zhang, H.G., Zhu, Q.S., Zhang, Y., Wang, Y., Zhao, L., Yu, B. (2007). One-pot synthesis and hierarchical assembly of hollow Cu_2O microspheres with nanocrystalscomposed porous multishell and their gas-sensing properties, *Adv Funct Mater, Volume 17,* 2766–2771.
[13] Wang, W.Z, Huang, X.W., Wu, S., Zhou, Y.X., Wang, L.J., Shi, H.L., Liang, Y.J., Zou, B. (2013). Preparation of p-n junction $Cu_2O/BiVO_4$ heterogeneous nanostructures with enhanced visible-light photocatalytic activity, *Appl Catal B Environ, Volume 134,* 293–301.
[14] Liu, Y.B., Zhou, H.B., Li, J.H., Chen, H.C., Li, D., Zhou, B.X., Cai, W.M. (2010). Enhanced photoelectrochemical properties of Cu_2O-loaded short TiO_2 nanotube array electrode prepared by sonoelectrochemical deposition, *Nano-Micro Lett, Volume 2,* 277–284.
[15] Kuo, C.H., Huang, M.H. (2010). Morphologically controlled synthesis of Cu_2O nanocrystals and their properties, *Nano Today, Volume 5,* 106–116.
[16] Zhang, X., Yu, S., Cao, Y., Chen, L., Zhao, T. and Yang, F. (2013). Cu_2O-coated polystyrene microsphere materials with enhanced photo- and photoelectro-catalytic activity, *J. Solid State Electrochem., Volume 17,* 1429–1434.
[17] Basu, M., Sinha, A. K., Pradhan, M., Sarkar, S., Pal, A., Mondal, C. and Pal, T. (2012). Methylene Blue−Cu_2O Reaction Made Easy in Acidic Medium, *J. Phys. Chem. C, Volume 116,* 25741−25747.
[18] Xu, Y., Jiao, X., Chen, D. (2008). PEG-assisted preparation of single-crystalline Cu_2O hollow nanocubes, *J Phys Chem, Volume 112,* 16769–16773.
[19] Teo, J.J., Chang, Y., Zeng, H.C. (2006). Fabrications of hollow nanocubes of Cu_2O and Cu via reductive self-assembly of CuO nanocrystals, *Langmuir, Volume 22,* 7369–7377.
[20] Zhang, H., Zhang, X., Li, H., Qu, Z., Fan, S., Ji, M. (2007). Hierarchical growth of Cu_2O double tower-tip-like nanostructures in water/oil microemulsion, *Cryst Growth Des, Volume 7,* 820–824.
[21] Kim, M.H., Lim, B., Lee, E.P., Xia, Y.J. (2008). Polyol synthesis of Cu_2O nanoparticles: use of chloride to promote the formation of a cubic morphology, *J Mater Chem, Volume 18,* 4069–4073.
[22] Zhang, H., Ren, X., Cui, Z. (2007). Shape-controlled synthesis of Cu_2O nanocrystals assisted by PVP and application as catalyst for synthesis of carbon nanofibers, *J Cryst Growth, Volume 304,* 206–210.
[23] Huang, L., Peng, F., Yu, H., Wang, H. (2009). Preparation of cuprous oxides with different sizes and their behaviors of adsorption, visible-light driven photocatalysis and photocorrosion, *Solid State Sci, Volume 11,* 129–138.
[24] Kuo, C.H., Huang, M.H. (2008). Fabrication of truncated rhombic dodecahedral Cu_2O nanocages and nanoframes by particle aggregation and acidic etching, *J Am Chem Soc, Volume 130,* 12815–12820.
[25] Zhou, Z.H., Lin, Y.L., Zhang, P.G., Ashalley, E., Shafa, M., Li, H.D., Wu, J., Wang, Z.M. (2014). Hydrothermal fabrication of porous MoS_2 and its visible light photocatalytic properties, *Mater Lett, Volume 131,* 122–124.

[26] Hu, P., Yu, L.J., Zuo, A.H., Guo, C.Y., Yuan, F.L. (2009). Fabrication of monodisperse magnetite hollow spheres, *J Phys Chem, Volume 113*, 900–906.
[27] Siegfried, M.J., Choi, K.S. (2005). Directing the architecture of cuprous oxide crystals during electrochemical growth, *Angew Chem, Volume 117*, 3282–3287.
[28] He, P., Shen, X., Gao, H. (2005). Size-controlled preparation of Cu_2O octahedron nanocrystals and studies on their optical absorption, *J Colloid Interface Sci, Volume 284*, 510–515.
[29] Chen, Q., Shen, X., Gao, H. (2007). Formation of solid and hollow cuprous oxide nanocubes in water-in-oil microemulsions controlled by the yield of hydrated electrons, *J Colloid Interface Sci, Volume 312*, 272–278.
[30] El hajji, M., Hallaoui, A., Bazzi, L., Benlhachemi, A., Jbara, O., Tara, A., Bakiz, B. (2014). Nanostructured ZnO, ZnO-CeO_2, ZnO-Cu_2O Thin Films Electrodes Prepared by Electrodeposition for Electrochemical Degradation of Dye, *Int. J. Electrochem Sci, Volume 9*, 4297-4314.
[31] Yang, H., and Liu, Z. (2010), Facile Synthesis, Shape Evolution and Photocatalytic Activity of Truncated Cuprous Oxide Octahedron Microcrystals with Hollows, *Crystal Growth & Design, Volume 10,* 2064-2067.
[32] Ji, F., Li, C., Zhang, J. and Deng, L. (2011). Heterogeneous photo-Fenton decolorization of methylene blue over $LiFe(WO_4)_2$ catalyst, *Journal of Hazardous Materials, Volume 186,* 1979–1984.

SOLUTION PRECURSOR PLASMA SPRAYED SUPERHYDROPHOBIC SURFACE

Yuxuan Cai[1], Gisele Azimi[2], Thomas W. Coyle[1], and Javad Mostaghimi[1]
[1]Centre for Advanced Coating Technologies (CACT), Department of Mechanical & Industrial Engineering, University of Toronto, Toronto, ON, Canada
[2]Department of Chemical Engineering & Applied Chemistry, University of Toronto, Toronto, ON, Canada

ABSTRACT

This work presents the first rare earth metal oxide superhydrophobic surfaces fabricated by using the solution precursor plasma spraying technique. The water contact angle measured on the surface is 161°. The dynamic impact of a water droplet is also captured by a high speed camera. The topography and cross-sectional microstructures of the coated surface are examined by scanning electron microscopy

INTRODUCTION

Since the development of electron microscopy in the last century, people have been able to study nano-textures of plant surfaces and animal skins. It has been found that a unique surface architecture provides a self-cleaning ability to leaves (e.g. aspen leaf), which is shown in Fig 1, and allows the wings of insects to remain dirt-free and shed water when they fly in the rain[1]. This type of surface is called a hydrophobic surface. Hydrophobic surfaces are characterized by a high water contact angle (>90°) and a low roll-off angle. When the water contact angle on the surface is higher than 150°, the surface is called superhydrophobic. Due to these properties, hydrophobic surfaces have a wide range of potential applications to benefit the environment in energy conservation, including enhancement of condensation in steam power plants in order to increase the efficiency of electricity generation[2]; and promotion of nucleation for pool boiling at low heat flux to enhance boiling heat transfer[3]. Very recently, rare earth oxides (REOs) have been proposed as a means of creating hydrophobic surfaces[2]. Due to their unique electron configuration and high melting temperature (over 1500°C), rare earth oxide surfaces exhibit a high water contact angle even when they are smooth and at elevated temperatures. However, the link between this research and industrial applications has not been firmly established due to the difficulties of large scale fabrication, and complex processing procedures.

Plasma spray deposition is a technique which has been widely used in industry to produce coatings due to its high deposition rate, near-net shape finishing, and most importantly, its ability to process almost all materials. Conventional plasma spray uses micro scale powder particles as the feedstock for the material to be deposited. Solution precursor plasma spray (SPPS) is a relatively new technique that, rather than using powder particles, uses a solution which decomposes during deposition to form the coating. The coating formation mechanisms for solution precursor plasma spray are different than for conventional plasma spray, and lead to nano- and submicron-structured coatings[5]. This type of hierarchical structured surface (i.e. a surface consists of both nano- and submicron-scale features) is very desirable in fabricating hydrophobic surfaces since it captures the essence of self-cleaning leaves and wings in nature. To our knowledge, there has yet to be a superhydrophobic surface produced by the solution plasma spray method.

In this work, superhydrophobic coatings were fabricated using axial-injection SPPS. Rare earth metal salt dissolved in the water was used as the liquid feedstock to the plasma torch. The wetting behavior of the coated surface is characterized by measuring the static water contact angle, and the dynamic impacts of water droplets on the surface. Surface and cross-sectional microstructures were analyzed by scanning electron microscopy (SEM).

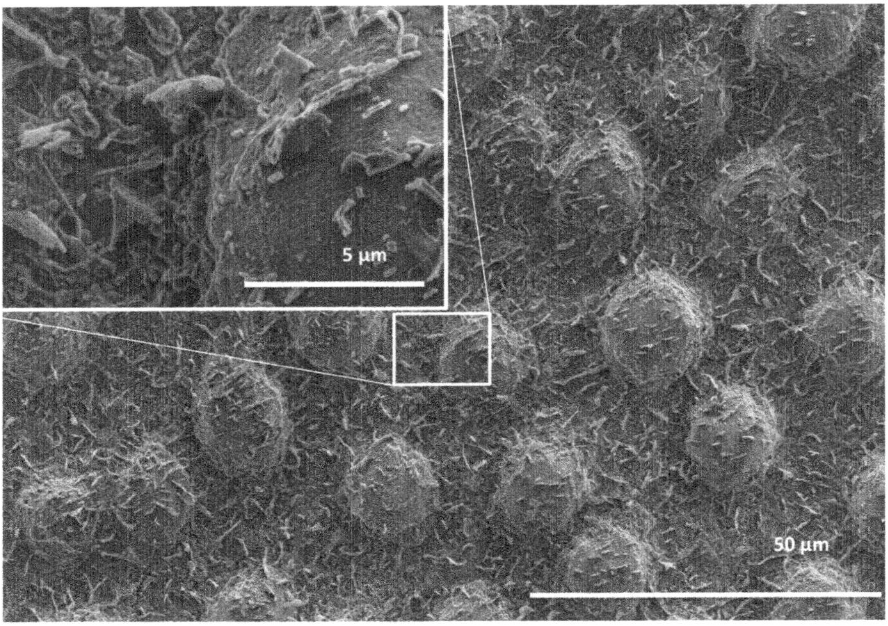

Figure 1: Scanning electron microscopy (SEM) image of an aspen leaf shows the hierarchical surface structure consisting of micro-scale clusters and nano features.

EXPERIMENTAL

Substrates of Type 304 stainless steel, 25.4 mm in diameter and 3.175 mm in thickness, were coated with Yb_2O_3 by the SPPS technique. The substrates were fabricated from cold-rolled steel by the supplier, and used with the surface finish as-received from the supplier. Prior to the plasma spray deposition, the substrates were pre-heated to 350°C. Ytterbium oxide was selected as the coating material from the REOs due to the high solubility of ytterbium nitrate in water and economic consideration.

The coating solution was prepared at room temperature by dissolving 99.999% Ytterbium nitrate pentahydrate (Pangea International, Shanghai, China) in distilled water. The concentration of the solution was 167.3 g of ytterbium nitrate per 100 g of water. The feedstock rate was set to 15 g/min. An Axial III Series 600 plasma torch (Northwest Mettech Corp., North Vancouver, BC, Canada) was used to deposit the coating by the SPPS process onto the substrates. The nozzle size was 9.525 mm in diameter. Argon was used as the atomizing gas at a flow rate of 10 slpm. The plasma gas consisted of a mixture of hydrogen (10%), nitrogen (80%) and argon (10%) was used at a flow rate of 250 slpm. The current applied in the experiment was 200 A per set of electrodes, and the torch power was approximately 125 kW. The standoff distance was set at 80 mm. A robot arm which carried the torch moved in a raster pattern, and in total 15 passes were performed on each substrate. The robot arm was set to a linear translation speed of 85 mm/s and the vertical step size was 5 mm.

The as-sprayed coatings were sectioned by a precision diamond saw, IsoMet 5000 (Buehler, ON, Canada). The sectioned samples were then mounted in epoxy under a vacuum of 3000 Pa. A low viscosity epoxy Jetset Epoxy (MetLab Corp, ON, Canada) was selected to allow the epoxy to penetrate into the pores of the cross section of the coating. The mounted coating's cross section was subsequently polished using P320 silica grinding paper, followed in sequence by 45 μm, 15 μm, 6 μm, and 3 μm diamond disks. Then 1 μm and 0.05 μm diamond suspensions were used as the final stages of polishing. Between each polishing step, the surface was cleaned in an ultrasonic bath and dried by compressed air. Scanning electron microscopy (Hitachi SU 3500) was used to characterize the surface and cross-sectional microstructures. To avoid charging effects in the SEM, specimens of the top surfaces of the coatings were sputtered coated with gold and the polished cross sections were coated with carbon before observation.

The static water contact angle image was captured by a CCD camera (Sony XCD-SX900) with a horizontal microscope (Wild Heerbrugg 400076) at 5.8x magnification. The droplet was illuminated by a white-light projector from behind through a frosted glass. Image processing software (ADSA) was used to analyze the image of the static water droplet in order to determine the static water contact angle. The dynamic impacts of water droplets on the coating were captured by a high speed camera FASTCAM SA5 (Photron, CA, USA) at 4000 frames per second.

X-ray diffraction (XRD) was performed to study the phase of the coating material using a Miniflex600 (Rigaku, MI, USA). The coating was cut into 15 mm by 15 mm square specimens in order to fit the sample holder of the XRD machine. The measurements were performed over a range of 2θ angle from 15° to 105°, and the patterns obtained were compared with standard reference patterns.

Results and discussion

Surface and Cross-sectional Microstructures

Examining the microstructure of the surface of the coating, a hierarchical microstructure was observed as expected. Figure 2 shows the SEM image of the surface. Micro-scale irregular clusters ranging from 5 micron to 10 micron in size were uniformly formed on the surface. On the clusters, nanoscale particles were also observed.

In SPPS, the coating is in general formed by the stacking of irregular particles and molten splats, therefore, micro-scale agglomerates are commonly observed. The nanoscale particles may have resulted from the condensation of vaporized particles. Another possibility is the nano-particles passed through the boundary layer due to the thermophoresis force. Notice that this type of microstructure is very similar to the hierarchical structured hydrophobic surfaces in nature.

Figure 3 shows an SEM image of the cross-section. The coating is porous and a feathery structure is observed. The particles observed in the coating have irregular shapes which is an indication of incomplete melting. The coating appears to have been formed mainly by the sintering of un-melted (or re-solidified) particles. In order to understand the extent of melting of the deposition material, single torch pass depositions were also performed. The results of the single torch deposition showed deposits mainly consist of pancake shaped splats and more and more incompletely melted particles were observed at the periphery of the plasma plume. The plasma temperature is lower at the periphery which may be the reason for the incompletely melted material. Careful examination of the cross section images reveals that at the substrate-coating interface, evidence of pancake shaped splats can be seen, but only for the first few layers adjacent to the substrate. This may due to the incompletely melted particles do not adhere well to the bare substrate on the first torch pass, so that pancake shaped splats formed from the fully melted droplets are observed. For the subsequent torch passes, incompletely melted particles were able to adhere to the previously deposited splats and sintered during the subsequent torch passes. As the coating was built-up by the deposition of these particles, the porosity was enhanced by a shadowing effect which led to the formation of the feathery structure. The coating was denser close to the substrate than near the surface. This may be the result of the coating closer to the substrate experiencing more torch passes, and therefore more sintering, than the top of the coating.

Figure 2: Scanning electron microscope images of the top surface showing micro-scale clusters (a) and the fine nanoscale particles on the clusters (b).

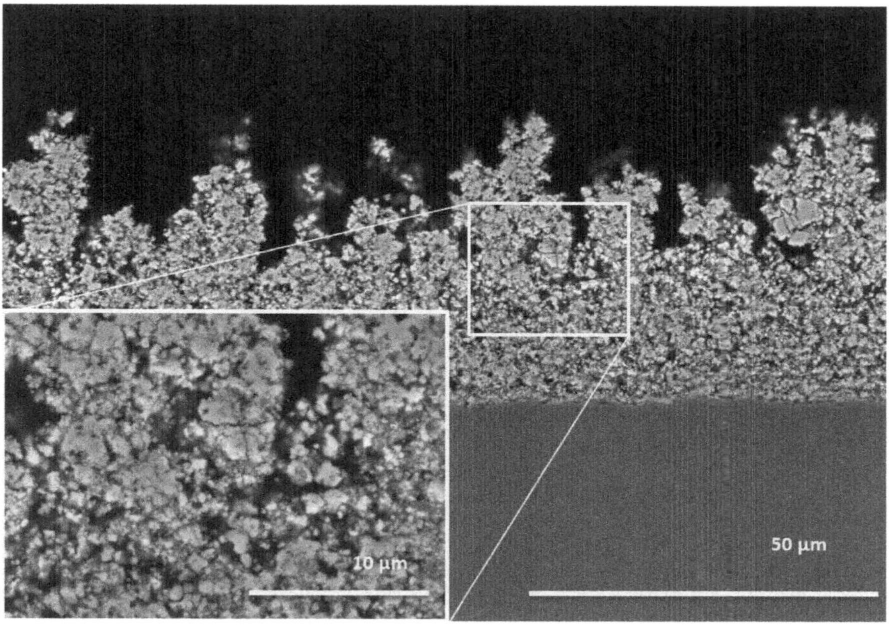

Figure 3: Scanning electron microscope images of the cross-section.

X-ray Diffraction

Figure 4 shows the XRD pattern of the plasma sprayed coating. The sharp peaks in the XRD patterns indicate that the coating material is crystalline. As shown in Figure 3, the peaks in the pattern match with the Yb_2O_3 reference (PDF card No.: 01-075-6635) at all peak locations. No additional peaks corresponding to the substrate are observed, which indicates the coating material fully covered the substrate.

Wetting Behavior of the Coating.

A water contact angle of 161°, which indicates the coating is superhydrophobic, was observed. On a smooth rare earth metal oxide surface, the measured water contact angle was 100°[2]. Thus, the hierarchical microstructure of the coating formed by the SPPS method enhanced the water contact angle over 60%. The roll-off angle, another important characteristic of the wetting behavior, is so small that the water droplet rolls off at the slightest tilt. Figure 5 shows the water droplets on the coating and the measured water contact angle.

The dynamic impact of a single droplet and the coalescence between two droplets were examined using a high speed camera. For a single droplet impacted on the coating at a speed of 1.4 m/s, the droplet completely rebound. When a second droplet landed beside a static droplet at the same velocity, it deformed and interacted with the first droplet. Then the two droplets combined into one and the combined droplet was also recoiled. The combination of the hierarchical surface structure and the intrinsic hydrophobic material gives the coating an excellent water repellent property. Figure 6 shows the dynamic impacts of water droplets on the as-sprayed coating. The combination of the hierarchical surface structure and the intrinsically hydrophobic material gives the coating an excellent water repellent property.

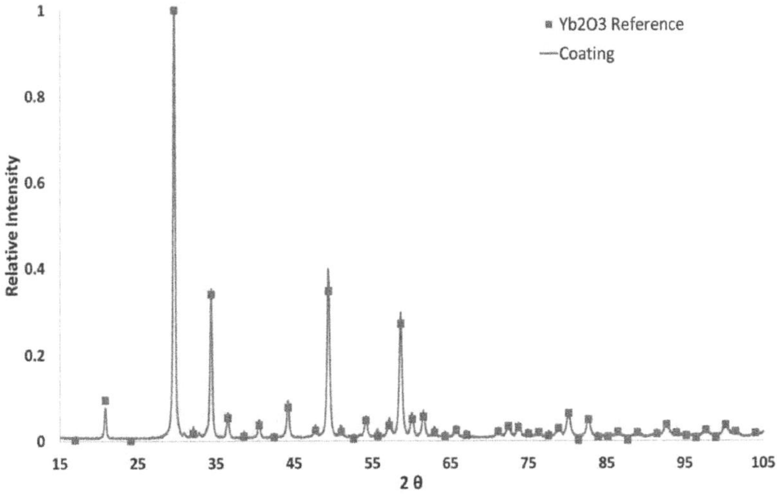

Figure 4: X-ray diffraction pattern of the plasma sprayed coating compared with the Yb_2O_3 reference (PDF card No.: 01-075-6635)

Figure 5: Water droplets on the coated surface. The droplet size is 2.5 mm.

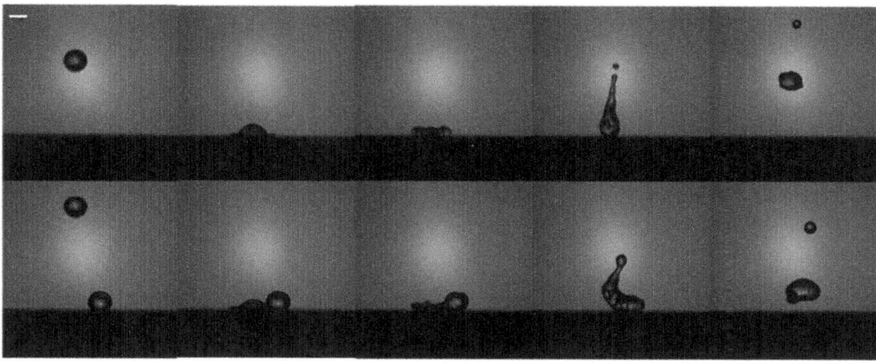

Figure 6: The dynamic impact of water droplets on the coated surface. Scale bar: 2 mm. Top: single droplet. Bottom: droplets coalescence.

CONCLUSIONS

This work shows a promising method of fabricating superhydrophobic coatings using precursor solutions as feedstock in a plasma spray deposition process. It offers a fast, simple and low cost method to fabricate a hydrophobic surface in a large scale. The coated surface has a hierarchical structure which is similar to lotus leaves. The water contact angle measured is 161° and the roll-off angle is less than 5°. The dynamic impact tests show a great water repellent property of the coating.

ACKNOWLEDEGMENTS

First, the authors would like to thank the support from our sponsor, NSERC. Also YC thank Dr. Larry Pershin, Dr. Jeffery Harris, Mr. Sal Boccia, and Mr. Tiegang Li for providing him the technical expertise required to develop and implement the SPPS idea of the work.

Furthermore, YC wants to thank Mr. Pedro Isaza, Mr. Alireza Dalili, Mr. Joel Castonguay, Mr. Gengxu Yan, and his mother for their supports and helps.

REFERENCES

[1] H. Ensikat, P. Ditsche-Kuru, and P. Neinhuis. Beilstein J. Nanotechnol, 152, 2 (2011)

[2] Azimi G, Dhiman R, Kwon H, Paxson A, Varanasi K. 2013. Hydrophobicity of rare-earth oxide ceramics. Nature Material 12: 315-320

[3] Betz AR, Jenkins J, Kim CJ, Attinger D. 2013. Boiling heat transfer on superhydrophilic, superhydrophobic, and superbiphilic surfaces. Int J Heat and Mass Transfer 57: 733-741

IMPROVEMENT OF INTERFACIAL STRENGTH FOR THERMAL BARRIER COATINGS BY FORMATION OF WEDGE-LIKE THERMALLY GROWN OXIDE

Kazuhiro Ogawa [1], Shun Hatta[1] and Hiroyuki Yamazaki[2]
[1] Fracture and Reliability Research Institute, Tohoku University, 6-6-11, Aoba, Aramaki, Aoba-ku, Sendai 980-8579, Japan
[2] Tohoku Electric Power Co., Inc., 1-7-1, Honcho, Aoba-ku, Sendai 980-8550, Japan

ABSTRACT

High-temperature coating systems, which were used in gas turbine engines, usually consist of a ceramic Thermal Barrier Coating (TBC) on top, an intermediate metallic-Bond Coating (BC) and a superalloy substrate. It is usually used that yttria-stabilized zirconia (YSZ) as a top coating (TC) and CoNiCrAlY alloy as bond coating, respectively. During service, a Thermally Grown Oxide (TGO) layer grows at the interface between the TC and the BC. The TGO is predominantly comprised of alpha-Al_2O_3 and provides protection of the underlying substrate against high-temperature corrosion. However, upon cooling of the component from high temperature, thermal-mismatch strains can be generated within the TGO. The elastic energy associated with these strains is the dominant driving force for failure of the coating systems. This strain energy increases with increasing TGO thickness. This explains that the failure of a high-temperature system can often be correlated with a critical TGO thickness at failure.

In this study, in order to control the TBC life, new bond coat materials with excellent interfacial strength for thermal barrier coatings was developed. By using Ce added bond coat material, wedge-like TGO was form at the interface between a top coat and a bond coat. As a result, interfacial strength was improved by the Ce added bond coat material.

INTRODUCTION

In recent years, due to the demand for saving energy and reducing greenhouse gases, high-efficiency power generation plants have been developed. In power generating gas turbine plants, improvements in efficiency have been achieved by increasing turbine inlet gas temperature. Hot section components, such as combustors, rotating blades and vanes, are exposed to high temperature environments for prolonged periods of time. Accordingly, the materials for these components require resistance to oxidation and sulfurization at high temperatures. Therefore, thermal barrier coatings (TBCs) are important components of advanced gas turbines that require high reliability and endurance[1, 2]. A typical duplex TBC consists of a 100 μm thick metallic MCrAlY (M is Ni, Co or a combination of both) alloy bond-coat (BC) and a 250-300 μm thick ceramic top-coat (TC). The BCs, which function to improve bonding strength and high temperature corrosion resistance, are mainly produced by low pressure plasma spraying (LPPS). Generally, partially yttria stabilized zirconia (YSZ) has been used for TCs, which are mainly produced by air plasma spray (APS). When the turbine blades and combustors are exposed to high temperatures for long periods of time, it is highly probable that critical failure of TBCs, such as cracking, delamination or spalling, might occur. Improvement of bonding strength is one significant way to improve the reliability of TBCs over the long term.

During exposure to high temperatures, thermally grown oxide (TGO) forms at the interface between the TC and BC. The TGO layer of a plasma sprayed TBC consists of two layers, one is an Al_2O_3 layer, and the other is a mixed oxide layer that consists of NiO, CoO, Cr_2O_3, or $(Ni,Co)(Cr,Al)_2O_4$ spinel oxide[3]. It is generally accepted that the formation of the TGO accelerates TBC failures, and the bonding strength of the TBCs is reduced by the growth of the TGO[4, 5]. Ogawa et al. reported that the formation of pores in a mixed oxide caused stress concentration sites and a decrease in bonding strength3. Therefore, inhibition of TGO formation, especially the inhibition of mixed oxide formation, would be an effective means for improving bonding strength. It is reported that the TGO growth rate depends on chemical composition, existence of atomic-order defects in the BC, TC materials etc. [4, 6]. Some approaches, such as modification of the bond-coat or of the coating processes, along with optimization of coating conditions to inhibit the formation of TGO have been considered. We have developed a modified bond-coat material by adding Ce to a conventional material [7, 8]. The TGO morphology of the TBC with developed BC material showed completely different TGO compared with the TBC with conventional CoNiCrAlY bond coat material, namely wedge-like TGO (inward oxide). As a result of formation of the wedge-like TGO, we succeeded in producing a TBC with a bonding strength that is three times higher than conventional TBC. However, the wedge-like TGO was formed at more than 1100°C. It means that the wedge-like TGO cannot be formed at the usual service temperature of the gas turbines. Therefore, in this study, in order to generate the wedge-like TGO at usual service temperature around 1000°C, effects of pre-oxidation for a TBC with Ce-containing BC are investigated.

EXPERIMENTAL

MATERIALS USED AND SPRAYING SYSTEM

Polycrystalline Ni base superalloy, INCONEL738LC, was used as substrate materials. The substrate was initially grit blasted, and then approximately 100μm-thick bond coats were coated by cold spray (CS) system. Three kinds of feedstock powders of the BC materials were prepared as shown in Table I. Hereafter, conventional CoNiCrAl, and developed CoNiCrAlYCe and CoNiCrAlCe bond coat materials call in abbreviated words as “STD”, “YCe” and “CE”, respectively. The commercial CS facility PCS-800 (Plasma Giken Co., Ltd.) was used for cold sprayed bond coats. The facility used helium as the propellant gas. A de-Laval nozzle gun was used. And an yttria (8wt.%-Y_2O_3) stabilized zirconia (YSZ) TBC top coating, approximately 300μm-thick, was applied using an air plasma spraying (APS) coating system. The coating conditions are presented in Table II.

Table I: Chemical compositions of used BC. (wt.%)

BC materials	Co	Ni	Cr	Al	Y	Ce
CoNiCrAlY (STD)	Bal.	32	22	8	0.5	-
CoNiCrAlY+Ce (YCe)	Bal.	32	22	8	0.5	1.5
CoNiCrAl+Ce (CE)	Bal.	32	22	8	-	1.5

Table II: Spray conditions of used specimens.

Substrate		Ni base superalloy Inconel 738LC
CS gas condition		He 3.0MPa / 600°C
Top coat (300μm)	Powder	8 wt.% YSZ
	Construction technique	Air Plasma Spray (APS)
Bond coat (100μm)	Powder	CoNiCrAlY CoNiCrAlYCe CoNiCrAlCe
	Construction technique	Cold Spray (CS)

PRE-OXIDATION HEAT TREATMENTS AND HIGH TEMPERATURE EXPOSURE TESTS

High temperature exposure tests were performed at 1100oC for 20hours as pre-oxidation heat treatment and then at 1000°C for 1000hours in atmospheric environment inside a muffled furnace. TGO formation was investigated by scanning electron microscopy (SEM) and energy dispersive X-ray spectrometry (EDX) for TBCs with three different bond-coat materials. A schematic of heat treatments is shown in Figure 1.

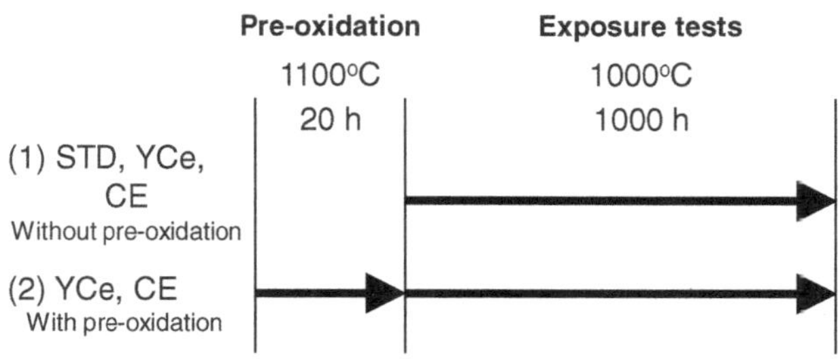

Figure 1. A schematic of heat treatments.

And also, in order to investigate the effect of cold spray conditions, the propellant gas conditions (gas pressure and temperature) were changed as follows.

1) Pressure: 3MPa, Temperature: 600°C
2) Pressure: 4MPa, Temperature: 600°C
3) Pressure: 4MPa, Temperature: 800°C

FOUR-POINT BENDING TESTS

Prior to four-point bending tests, specimens, which have dimensions of 50 mm × 5 mm × 3.4mm, were aged at 1100°C for 20hours as pre-oxidation heat treatment and then for 1000hours at 1000°C. As shown in Figure 2, during the test, an acoustic emission (AE) sensor and a strain gauge measured AE energy and the compressive strain on the specimens' backside, respectively. Assuming that the starting point of a rapid increase in cumulative AE energy corresponds to the occurrence of delamination, bonding strength was evaluated quantitatively by the amount of backside strain at the point of a rapid increase in cumulative AE energy.

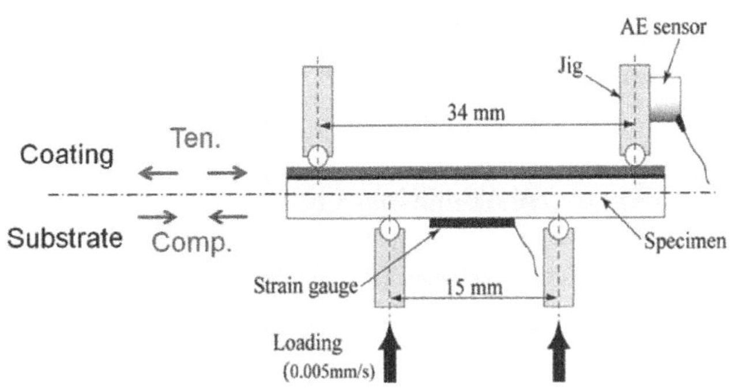

Figure 2. Schematic illustration of four-point bending test .

RESULTS AND DISCUSSION

SEM OBSERVATION

Typical SEM images of as-sprayed TBC specimens are shown in Figure 3. As shown in the figure, dense bond coat layers were obtained by cold spray technique.

Typical SEM images of pre-oxidation heat treated TBC specimens are shown in Figure 4.

In the case of STD specimen, delamination was occurred by only pre-oxidation heat treatment at 1100°C for 20hours. Therefore, there is no STD image in Figure 4.

After pre-oxidation heat treatment at 1100°C for 20hours, inward wedge-like TGOs were formed at the interface between TC and BC for YCe and CE specimens (see Figure 4). Especially, in CE specimen, predominant wedge-like TGO was observed. The reason of formation of the predominant wedge-like TGO in CE specimen have not been made clear yet. It is an issue in the future.

In Figure 4, there are many white dots inside the black TGO layer. As a result of EDX analysis, the white dots are CeO_2. And black layer in SEM image is Al_2O_3, and grey layer is Cr_2O_3. From this result, it is thought that Ce has affected internal diffusion (formation of inward wedge-like TGO).

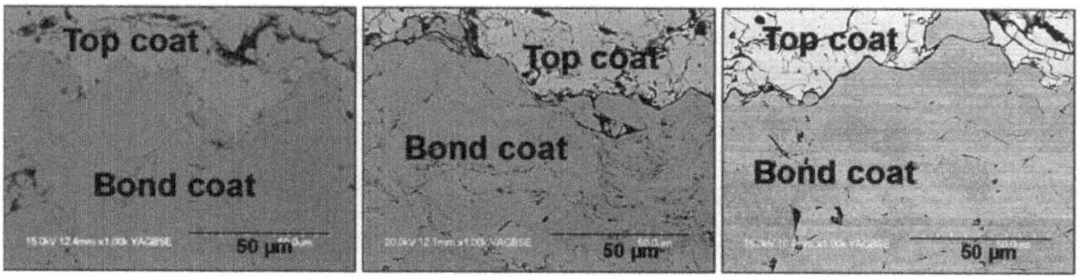

a) STD b) YCe c) CE

Figure 3. Cross-sectional SEM images of as-sprayed TBCs.

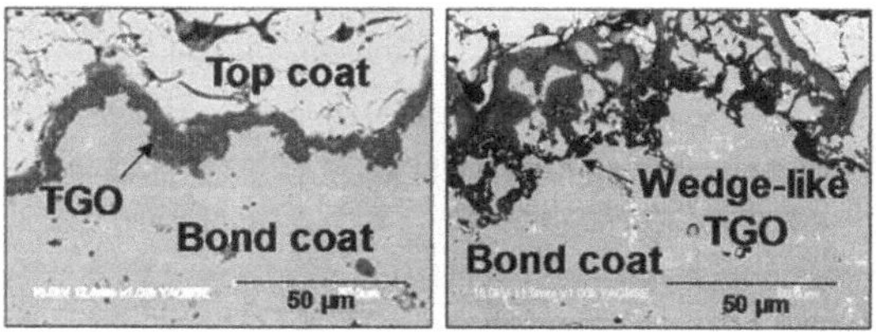

a) YCe b) CE

Figure 4. Cross-sectional SEM images of pre-oxidation heat treated TBCs at 1100°C for 20hours.

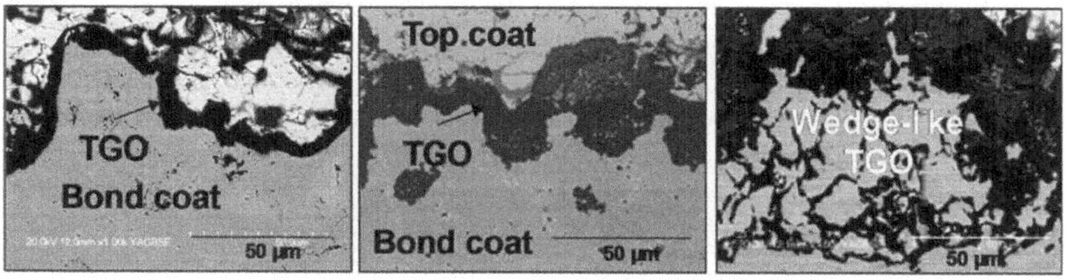

a) STD b) YCe c) CE

Figure 5. Cross-sectional SEM images of high temperature exposed TBCs at 1000oC for 1000hours.

Typical SEM images of high temperature exposed TBC specimens are shown in Figure 5.

After high temperature exposure test, these wedge-like TGOs grew in YCe and CE specimens. On the other hand, in the STD specimen, an outward TGO was formed, and then grew. These TGO morphologies were quite different, and the TBCs with Ce content bond coats have thicker TGO than the TBC with conventional CoNiCrAlY.

OXIDE GROWTH

The thickness of the wedge-like TGO was measured at random 30 points and was averaged. The results of thickness measurements is shown in Figure 6.

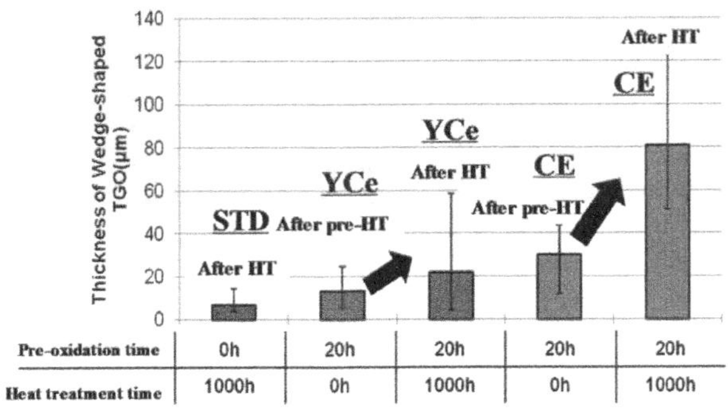

Figure 6. Results of TGO thickness measurements.

From this figure, by adding Ce to the conventional bond coat materials, the wedge-like TGOs were formed by pre-oxidation and were obviously grown by high temperature exposure. In the case of the CE specimen, more than 80μm-thick TGO was observed by high temperature exposure at 1000oC for 1000hours.

Effects of spray conditions on formation of the wedge-like TGO were investigated. As shown in Figure 7, the cold spray condition can vary wedge-like TGO formation behaviour. This means that it can be possible to control the wedge-like TGO formation by variation of the CS condition.

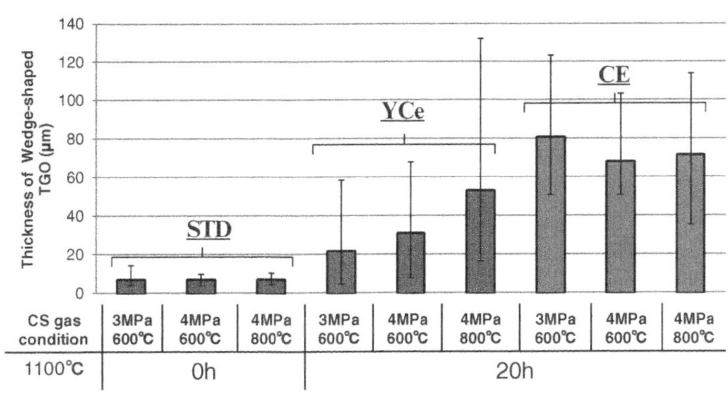

Figure 7. Effects of spray conditions on formation of the wedge-like TGO.

FOUR POINT BENDING TESTS

Results of four point bending tests for the TBCs without and with pre-oxidation heat treatments are depicted in Figure 8 and Figure 9, respectively.

In the case of the TBCs without pre-oxidation heat treatments, although spray condition is varied and Ce is added, there is a null effect. On the other hand, in the case of the TBCs with pre-oxidation heat treatments, delamination resistant property of YCe and CE was obviously improved by pre-oxidation. And also, the CS condition can affect interface strength. This results means that it can be possible to control the interface strength by variation of the CS condition.

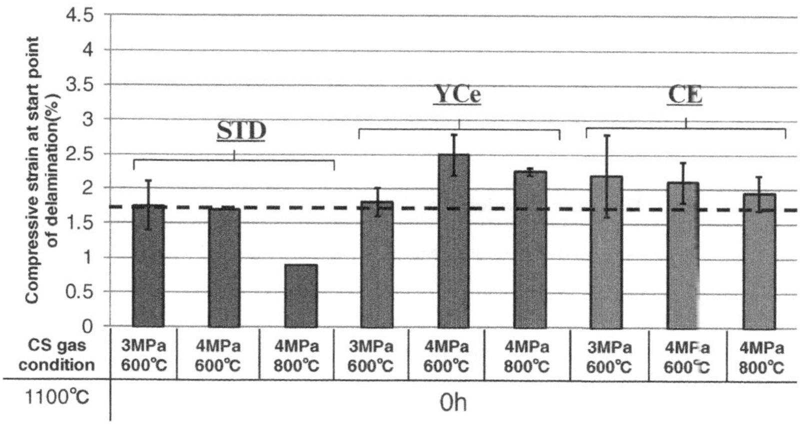

Figure 8. Results of four point bending tests for the TBCs without pre-oxidation heat treatments

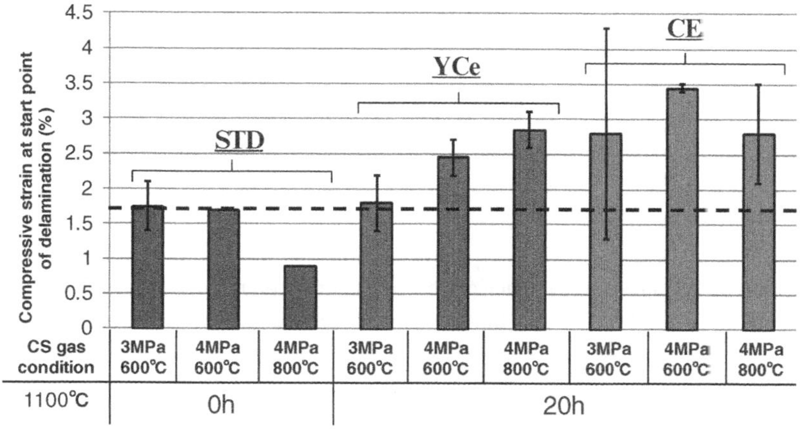

Figure 9. Results of four point bending tests for the TBCs with pre-oxidation heat treatments

Relationship between interface strength and the wedge-like TGO thickness is indicated in Figure10. From this result, interface strength increases with increasing thickness of the wedge-like TGO. Typically, growth of the TGO has bad influence for coating delamination, because the thick TGO can induce large internal stress. However, surprisingly, in the case of formation of wedge-like TGO, growth of the TGO has good influence for improvement of interface strength.

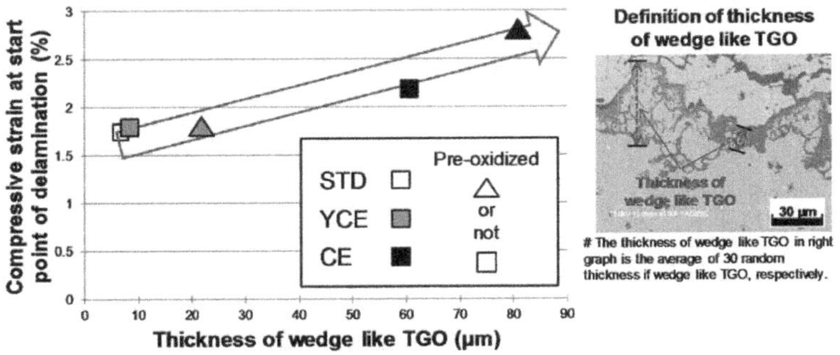

Figure 10. Relationship between interface strength and thickness of the wedge-like TGO.

CONCLUSIONS

In order to improve the interface strength between TBC and bond coat, effect of Ce content bond coat materials and effect of pre-oxidation heat treatments for the TBC with Ce content bond coats were investigated. The following main conclusions can be drawn:

(1) Wedge-like TGO was formed under 1000°C after pre-oxidation heat treatment.

(2) The interface strength was improved by pre-oxidation heat treatment.

(3) It can be possible to improve the interface strength by optimization of the CS condition.

ACKNOWLEDGMENT

This work was partly supported by the JSPS Core-to-Core Program, A. Advanced Research Networks,"International research core on smart layered materials and structures for energy saving".

REFERENCES

[1] S.M.Meier, D.K.Gupta, The Evolution of Thermal Barrier Coatings in Gas Turbine Engine Applications, Journal of Engineering for Gas Turbines and Power, Vol. 116, pp.250-257, 1994.

[2] B.J.Gill, R.C.Tucker, Jr., Plasma Spray Coating Process Materials Science and Technology, Vol.2, pp.207-213, 1986.

[3] K.Ogawa, T.Shoji, H.Aoki, N.Fujita, T.Torigoe, Mechanistic Understanding for Degraded Thermal Barrier Coatings, JSME int.J, Vol. 44A, pp. 507-513, 2001.

[4] G.M.Newaz, S.Q.Nusier, Z.A.Chaudhury, Damage Accumulation Mechanisms in Thermal Barrier Coatings, Journal of Engineering Materials and Technology, Vol. 120, pp.149-153, 1998.
[5] K.Ogawa, T.Kato, T.Shoji, Improvement of Interface Bonding strength Between Ceramic Thermal Barrier Coatings and Metallic Bond-coats, Proceedings of the International Thermal Spray Conference 2002, Erich Lugscheider, Ed., March 4-6, (Essen, Germany), pp. 900-904, 2002.
[6] H.Choi, B.Yoon, H.Kim, C.Lee, Isothermal oxidation of air plasma spray NiCrAlY bond-coats, Surface & Coatings Technology, Vol. 150, pp. 297-308, 2002.
[7] M.Tanno, K.Ogawa, T.Shoji, Effect of Cerium and Silicon Additions to MCrAlY on the High-Temperature Oxidation Behavior and Bond Strength of Thermal Barrier Coatings, Key Engineering Materials, Vols.261-263, pp.1061-1066, 2004.
[8] K.Ogawa, K.Ito, T.Shoji, D.W.Seo, H.Tezuka, H.Kato, Effects of Ce and Si Addition to CoNiCrAlY Bond Coating Material on Oxidation Behavior and Bonding Strength of Thermal Barrier Coatings, Journal of Thermal Spray Technology, Vol.15, pp.640-651, 2006.

EXPERIMENTAL PRODUCTION OF INDUSTRIAL ROLLER COATED BY HARD α-Al_2O_3 FILM USING AEROSOL DEPOSITION PROCESS

Naoki Seto[1], Kazuteru Endo[2], Noriaki Honda[2], Nobuo Sakamoto[1], Shingo Hirose[1] and Jun Akedo[1]
[1] National Institute of Advanced Industrial Science and Technology, 1-2-1, Namiki, Tsukuba, Ibraki, 305-8564, Japan
[2] Honda Seiki Co. ltd., 3-35-1, Tsurushirocho, Wakabayashi-ku, Sendai, Miyagi, 983-0034, Japan

ABSTRACT

Many mechanical parts, such as industrial rollers, shafts etc. is always required harder and smoother surface to achieve high efficiency production. These parts are usually coated by hard film such as hard Cr plating film. However, harder coating is required than hard Cr plating film for the industrial machine now. Therefore, the harder and the smoother film than Cr plating film is required strongly because such a hard film can improve mechanical operation efficiency by reducing maintenance time. On the other hands, α-Al_2O_3 film is one of the hardest material. And aerosol deposition (AD) method can coat α-Al_2O_3 film easily in room temperature.

In this paper therefore, α-Al_2O_3 film formation on industrial roll was developed using AD method. And various spec of α-Al_2O_3 film was investigated comparing with conventional hard Cr plating film.

INTRODUCTION

To achieve high efficiency production, the industrial machines are required precious action in long time with less maintenance. Therefore many mechanical parts of these machines, such as rollers, cams or shafts etc. are always required harder and smoother surface. At present, one of the most famous process to give hard and smooth surface to these mechanical parts is hard Cr plating film coating[1]. This hard Cr plating film can form comparatively easy and cheap. However, hard Cr plating film worn easier than hard ceramic film because this film is metal film. Therefore hard Cr plating film coating part's surface is injured soon in heavy working machine[2]. Thus, the harder and the smoother film than Cr plating (such as α-Al_2O_3) film is required strongly because such a hard film can improve mechanical operation efficiency by reducing maintenance time.

On the other hands, α-Al_2O_3 film can coat easily in room temperature using aerosol deposition (AD) method[3]. Therefore the authors tried to form α-Al_2O_3 film using AD method because this method is simple and easy process for hard α-Al_2O_3 coating. Moreover, evaluation of wear, chemical stability or friction coefficient of film surface were carried out by comparing between hard Cr plating and α-Al_2O_3 film. Furthermore, the authors were carried out experimental production of α-Al_2O_3 coating on industrial roller using modified AD method to consider industrial application of α-Al_2O_3 film or AD method.

EXPERIMENTAL SETUP

In α-Al_2O_3 film formation test, modified aerosol deposition chamber was used. Two type of setting (for substrate roll and for substrate plate) were used in this study. Fig. 1 shows these settings. To perform large area coating of α-Al_2O_3 using AD method, the authors modified chamber size or layout of chamber by putting rotation zig or X-Y axis arms for scanning large area. And Fine α-Al_2O_3 powder (about 1 micron diameter) was put in aerosol generator and supplied as coating material by carrier gas which was controlled by mass flow controller.

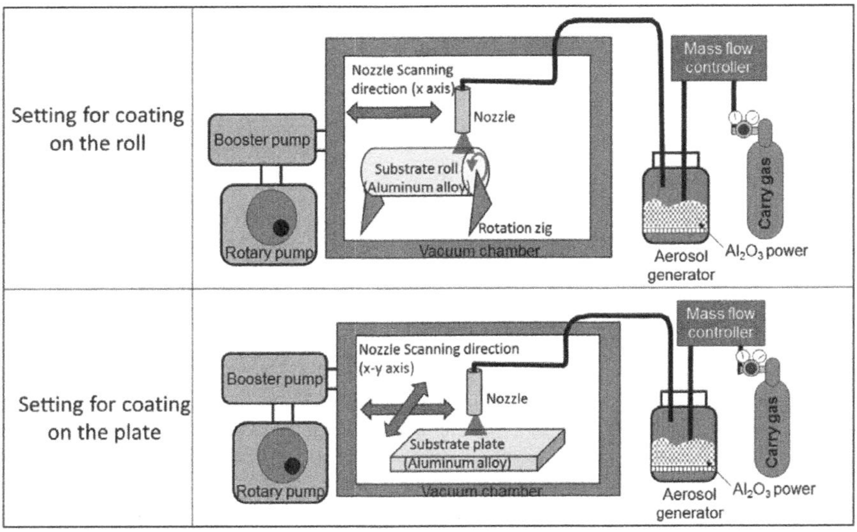

Fig. 1 Schematic drawing of aerosol deposition machine used in this study

After coating on the substrate, investigation or evaluation of Al_2O_3 film property was performed by SEM, XRD or Vickers hardness test. And to compare conventional hard coating, wear property and chemical stability were investigated by comparing between Al_2O_3 film and hard Cr plating film. Moreover friction coefficient of film surface was also investigated to consider Al_2O_3 film is really wear-less than hard Cr plating film.

RESULT AND DISCUSSION

Fig. 2 shows one of typical result of Al_2O_3 film coating on the industrial roll using AD method. We succeeded Al_2O_3 film on real industrial roller surface and the film thickness was 5 micro meter. The electrical resistance measurement between film surface and roller metal, the resistance was enough big to stop the electric current. This result indicates our Al_2O_3 film was coated successfully because Al_2O_3 is electrical insulator.

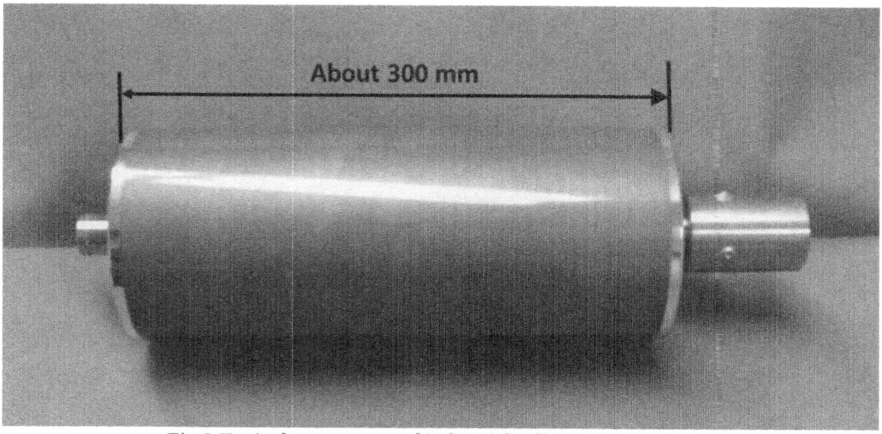

Fig.2 Typical appearance of industrial roller with Al_2O_3 film

To investigate this film was α-Al_2O_3 or not, SEM, XRD or Vickers hardness test were carried out. The surface observation of this film by SEM was smooth and dense. And observed peaks were only α-Al_2O_3 by XRD analysis. On the other hands, hardness of this film was measured. Vickers hardness test was performed in various point of this film, and the values were between 1100 and 1590 Hv. From these results, this Al_2O_3 film was fine and hard α-Al_2O_3 film, because the Vickers hardness of bulk α-Al_2O_3 is about 1450Hv. From these results, this Al_2O_3 film is considered fine and hard α-Al_2O_3 film.

If this Al_2O_3 film is fine and hard α-Al_2O_3 film, wear test result should be better than hard Cr plating film. Therefore wear property and chemical stability tests were carried out and investigated these abilities by comparing between hard Cr plating film and Al_2O_3 film.

Wear test was performed by ball-on disc examination. In this wear test, material of ball was Cemented Carbide, load of the ball was 500g (4.9N), Scratching speed was 78.5mm/sec and scratching time was 300 sec. Wear test was performed on Al_2O_3 film, hard Cr plating film and aluminum alloy(no film formation substrate). Fig. 4 shows wear graph and wear test track photo of Al_2O_3, Cr plating and aluminum alloy. According to this graph, Al_2O_3 was 4 times stronger than hard Cr plating film[4]. This result shows Al_2O_3 film has enough strong for wear-proof performance, and Al_2O_3 film can take the place of hard Cr plating film.

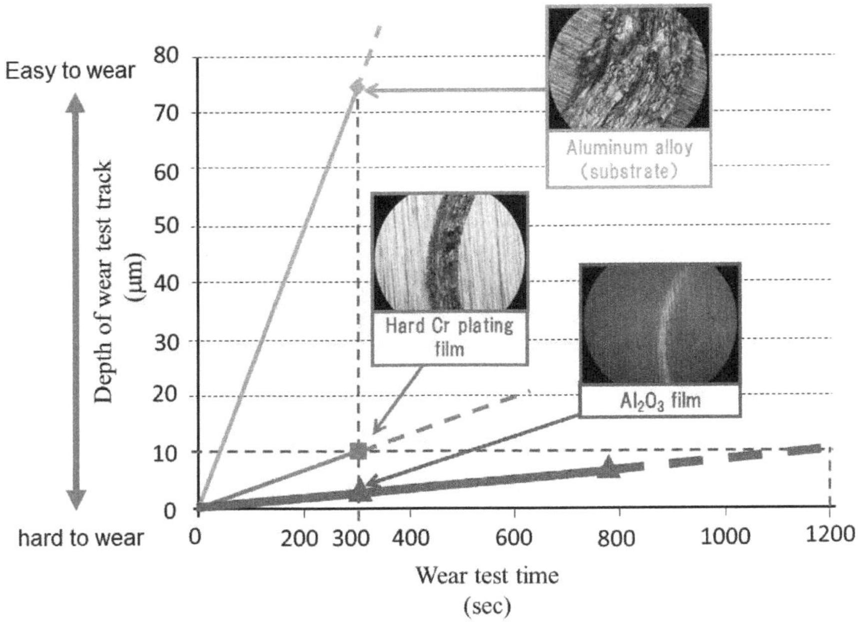

Fig. 4 Relationship between wear depth and wear test time

And chemical stability test was performed by neutral salt spraying test. Fig. 5 shows the salt spraying test result, and the condition of this test shows in Table 1. In Cr plating film, heavy dissolution or swelling could be observed. But in Al_2O_3 film, damage could not be observed. This result shows Al_2O_3 film is absolutely stable chemically than hard Cr plating film[4].

Table 1 Salt spray test conditions used in this study

Spray liquid	water (NaCl 50g/litter @pH6.5)
Testing cycle	15 cycles (Spray + Dry + Wet = 1cycle)
Test conditions	Spray (2 hours @ 308 K and 100% humidity) Dry (4 hours @333 K and 30% humidity) Wet (2 huors @ 323 K and over 95% humidity)

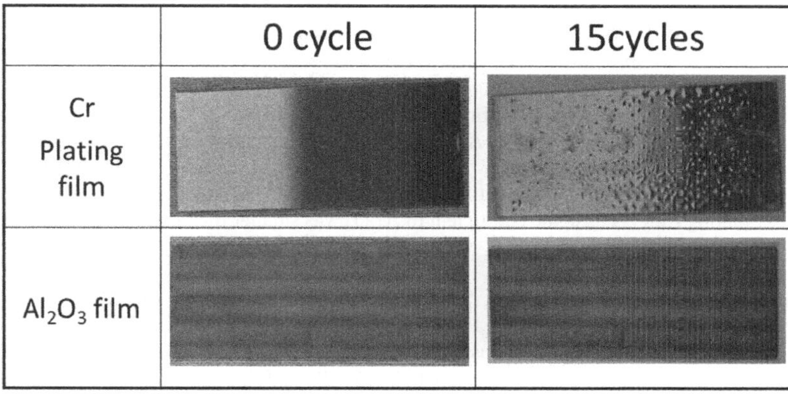

Fig. 5 Neutral salt spray test result of Al_2O_3 film and Cr plating film

Thus Al_2O_3 film coated by AD method was quite good results about wear-proof performance and chemical stability than hard Cr plating film. However, friction coefficient measurement was also important to evaluate our film is enough smooth and wear-less when this film applied to industrial parts. The friction coefficient of wear-proof surface is usually small because smooth surface is hard to wear. Therefore, friction coefficient of Al_2O_3 film was measured.

Fig. 6 shows photo of friction coefficient of Al_2O_3 film. In this measurement, friction coefficient can measured only the angle which the weight start to slide. This measurement was performed 3 times, and the average of our Al_2O_3 film's friction coefficient was 0.384.

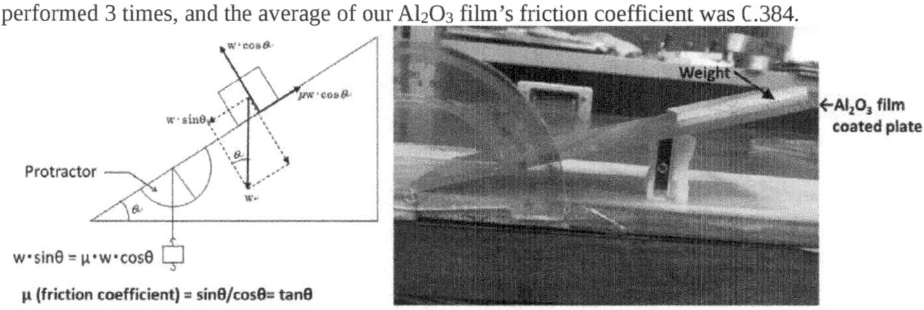

Fig. 6 Schematic drawing of friction coefficient measurement and photo of measurement

On the other hands, friction coefficient of hard Cr plating film was measured by DARBEIDA et. al., and its value was 0.79 [5]. This value was larger than our Al_2O_3 film's friction coefficient. This result indicates that our Al_2O_3 film has enough smooth and wear-less when this film applied to industrial parts because Al_2O_3 film has smaller friction coefficient than hard Cr plating film as conventional wear-proof film.

From these results, the authors success to coat hard and stable film on industrial roller or large plate. And Al_2O_3 film was stronger than conventional hard Cr plating film. From these results, industrial application of α-Al_2O_3 film coating on industrial mechanical parts by AD method may appear in the near future.

CONCLUSION

The authors tried to coat α-Al_2O_3 film using AD method and apply industrial roller coating instead of hard Cr plating film. Main conclusions are follows,

- ✓ We succeeded Al_2O_3 film on 300mm length real industrial roller surface.
- ✓ Our Al_2O_3 film was considered fine and hard α-Al_2O_3 film.
- ✓ In wear test and salt spray test, Al_2O_3 film was much strong and stable than conventional hard Cr plating film.
- ✓ According to friction coefficient measurement result, Al_2O_3 film was wear-less than hard Cr plating film because the coefficient of Al_2O_3 was smaller than hard Cr plating.

REFERENCES

[1] T. Arai, H. Fujita and M. Watanabe, "Evaluation of adhesion strength of thin hard coatings", Thin Solid Films, Vol. 154, Issues 1–2, P.387–401, 1987

[2] J.A. Picasa, A. Forna and G. Matthäusb, "HVOF coatings as an alternative to hard chrome for pistons and valves", Wear, Vol. 261, Issues 5–6, P.477–484, 2006

[3] J. Akedo, "Aerosol Deposition of Ceramic Thick Films at Room Temperature -Densification mechanism of ceramic layer –", JOURNAL OF THE AMERICAN CERAMIC SOCIETY, Vol. 89-[6], P.1834-1839, 2006

[4] N. Seto, K. Endo, N. Sakamoto, S. Hirose and J. Akedo, "Hard α-Al_2O_3 Film Coating on Industrial Roller Using Aerosol Deposition Method", Journal of Thermal Spray Technology, Vol. 23, Issue 8, p. 1373-1381, 2014

[5] DARBEIDA, A., STEBUT, J., BARTHOLE, M., BELLIARD, P., LELAIT, L., ZACHARIE, G. "Comparative tribological study of chromium coatings with differentspecific hardness" Surface & Coatings Technology 68/69, p. 582-590, 1994

Novel, Green, and Strategic Processing and Manufacturing Technologies

STEREOLITHOGRAPHIC ADDITIVE MANUFACTURING OF CERAMICS DENDRITES TO MODULATE ENERGY AND MATERIAL FLOWS

Soshu Kirihara
Joining and Welding Research Institute, Osaka University, 11-1 Mihogaoka Ibaraki 567-0047 Osaka, Japan

ABSTRACT

By using 3D printing process, various ceramics dendrites with spatially ordered micro cavities were successfully fabricated. Ceramic micro lattices with coordination numbers 4, 6, 8 and 12 were propagated spatially in computer graphic space. Ceramics nanoparticles were dispersed in to photo sensitive liquid resins at 40 % in volume fraction to obtain thixotropic slurries. The paste material was spread on a grass substrate with 10 μm in layer thickness by using a mechanical knife edge, and an ultra violet micro pattern was exposed to create cross sectional solid layer with 2 μm in part accuracy. After the layer stacking process, the obtained composite precursor was dewaxed and sintered in an air atmosphere. Through the computer aided design, manufacturing and evaluation, alumina micro photonic crystals with a diamond lattice structure were fabricated. Electromagnetic wave properties were measured by using a terahertz wave time domain spectroscopy. Subsequently, solid electrolyte dendrites of yttria stabilized zirconia with spatially ordered porous structures were successfully fabricated for fuel cell miniaturizations. Moreover, artificial bones of hydroxyapatite scaffolds with ordered porous structures were modeled successfully to realize excellent biological compatibilities.

INTRODUCTION

By an additive manufacturing of micro pattering stereolithography technique, novel electromagnetic devices of photonic crystals were fabricated. The artificial crystals with periodic arrangements of dielectric media can reflect the electromagnetic wave perfectly and can exhibit forbidden gaps in the transmission spectra by Bragg diffraction[1-3]. Through introductions of air cavities into the periodic lattices, the electromagnetic waves having specific wavelengths can resonate with structural defects, and localized modes of transmission peaks appear in the band gap. The alumina micro lattices with diamond structures were fabricated to control terahertz waves[4]. The terahertz waves are expected to detect micro cracks in material surfaces and structural defects in electric circuits by fine wave interferences, and to analyze cancer cells in human skins and toxic bacteria in natural foods through the higher frequency excitations[5-7]. The transmissions were measured by using a terahertz wave spectroscopy. The resonation behaviors were visualized by a transmission line modeling method of a finite difference time domain simulations. Subsequently, solid oxide fuel cells (SOFCs) are investigated as novel generation systems of electric powers with high efficiencies in energy conversion circulations. Yttria stabilized zirconia (YSZ) with high ion conductivities for incident oxygen is widely adopted material for solid electrolyte anodes as the SOFC components[8-10]. Recently, porous network structures were introduced into YSZ electrodes in micrometer or nanometer sizes to increase surface areas of reaction interfaces and gap volumes of stream paths[11-13]. In our investigation group, solid electrolyte dendrites composed of YSZ spatial lattice structures with various coordination numbers were fabricated successfully by the micro patterning stereolithography and powder sintering techniques to increase the surface areas of the electrode for fuel cell miniaturizations[14]. In the dendritic lattice structure, stress distributions and fluid flows were simulated and visualized by using finite element methods. These new computer aided designs, manufactures and evaluations have been established and optimized to create micro components of various ceramics in our investigation group. Moreover, artificial bones composed of hydroxyapatite (HAP) and beta-tricalcium phosphate (β-TCP) scaffolds with dendritic lattice structures were modeled successfully to realize excellent biological compatibilities[15]. Ordered porous structures were processed by using laser scanning stereolithography. The micro ceramics rods were arranged periodically to create lattice arrangements with four coordination number, and these aspect ratios were modulated continuously to realize biomimetic structures[16-19]. Fine microstructures without cracks or pores were observed by using a scanning electron microscope. These results show that the stereolithographic technique is a novel technology that can be used to fabricate bioscaffolds. The successfully manufactured bone substitute and the strictly designed bio-scaffold structures can be used to promote appropriate biochemical reactions and body fluid circulation in the near future through tissue engineering.

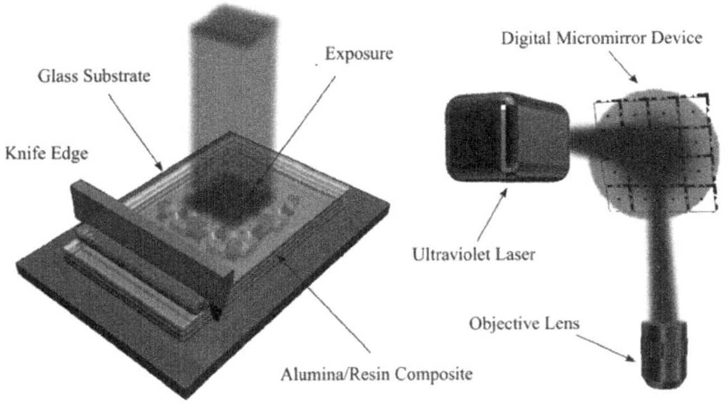

Fig. 1 Schematic illustration of 3D stereolithographic printer using photosensitive resin pastes with ceramic nanoparticle dispersions. Composite precursors were created by laminating 2D cross sections exposed by a digital micromirror device (DMD).

EXPERIMENTAL PROCEDURE

The micro pattering stereolithography system (D-MEC: SI-C1000, Japan) is illustrated schematically as shown in Fig. 1. The designed graphic data was converted into the stereolithographic format through polyhedral approximations. The solid model was sliced into the cross sectional numerical data sets to input the stereolithographic equipment. Photo sensitive acrylic resins dispersed with nanoparticles were fed over a substrate from a dispenser nozzle. The highly viscous resin paste was fed with controlled air pressure, and spread uniformly by a mechanical knife edge. The thickness of each layer was controlled to 10 μm. Two dimensional solid patterns were exposed on the slurry surface by a light induced photo polymerization. High resolution image had been achieved by using a digital micro mirror device. In this optical device, micro aluminum mirrors of 14 μm in edge length were assembled with 1024×768 in numbers. Each mirror could be tilted independently by piezoelectric actuating. The ultra violet lay of 405 nm was introduced into the digital micro mirror device, and the cross sectional image was reduced at 1/10 through an objective lens set and concentrated into the exposing area 1.3×1.7 mm in size. Through the layer stacking under the computer control, the acrylic resin component with the ceramics particles dispersion was obtained. Part accuracies of green bodies could be measured and observed by using a digital optical microscope (DOM). The formed models are dewaxed at 600 °C for 2 hs with a heating rate of 1.0 °C/min in the air atmosphere, and full ceramics components are obtained after sintering. Microstructures of the sintered components can be observed by using a scanning electron microscope (SEM). Relative densities of these ceramic components can be measured by Archimedian method.

The photonic crystals with the diamond lattices were designed by using the graphic application (Think Design: Toyota Caelum, Japan). The lattice constant was 1 mm in dimension. The whole structure was 6×6×2 mm^3 in size, consisting of 6×6×2 unit cells. The aspect ratio of the dielectric lattices was designed to be 1.5. The graphic data was transferred to the stereolithographic equipment. In the fabrication process, photo sensitive acrylic resin dispersed with alumina particles of 170 nm in diameter at 40 vol. % was used. The precursors with diamond structures were heated at various temperatures from 100 to 600 °C while the heating rate was set at 1.0 °C /min. The dewaxing process was observed in respect to the weight and color changes. Nanometer sized alumina particles could be sintered at 1500 °C with 8.0 °C/min in heating rate. The transmittance and phase shift of incident terahertz waves were measured by terahertz wave time domain spectroscopy (AISpec: J-Spec2001, Japan). Measured terahertz wave properties were compared with electromagnetic band diagrams theoretically calculated by a plane wave expansion method. Intensity profiles of the electric field were simulated by transmission line modeling (TLM) application (Flomerics: Micro Stripes, Germany).

The solid electrolyte dendrites with spatial lattice structures were designed by using the computer graphic application. The dendrite lattices of 1.16 in aspect ratio could be considered to exhibit the higher reaction efficiencies and gas transmittances according to Nernst equation. In the optimized dendrite structure, the diameter and length of YSZ rods were decided as 92 and 107 μm, respectively. The lattice constant was 250 μm. Photo sensitive acrylic resin dispersed with YSZ particles of 60 and 100 nm in first and second diameters at 30 vol. % were used for the stereolithography. The formed precursors with dendrite structures were heated at various temperatures from 100 to 600 °C while the heating rate was 1.0 °C /min. The dewaxing process was observed in respect to the weight and color changes. The YSZ particles could be sintered at 1500 °C for 2 hs. The heating rate was 8.0 °C/min. Fluid flow velocities and pressure stress distributions were simulated and visualized by finite volume method (FVM) application (Ansys: Cybernet Systems, Japan). In the simulations, no slip surface conditions were assumed.

The biological scaffolds with four coordinate rods were connected to form diamond lattice structures with a periodic length of 1.8 mm. The aspect ratios were adjusted in the range 1.2 to 2.0 in the scaffold models to control the porosity in the range from 50 to 80 vol. %. Photosensitive acrylic resins with the HAP or β-TCP particles at 40 vol. % were used for the materials of slurry. The fabricated precursors were dewaxed at 600 °C for 2 hs and sintered at 1250 °C for 2 hs in an air atmosphere. For the dewaxing and sintering, the heating rates were 1.0 and 5.0 °C/min, respectively. Fluid circulations were visualized with the fluid dynamic solver (ANSYS). Flow directions and velocities were calculated in spatial grids in the scaffold models through the finite element method (FEM). The simulation parameters were used to investigate the behavior of the blood flow in bone substitute. The fluid phase was represented as an incompressible Newtonian fluid with a viscosity of 1.45×10^{-3} Pa·s. The inlet velocity applied to the scaffolds was constant at 0.235 mm/s, and the pressure was zero at the outlet.

RESULTS AND DISCUSSION

The photonic crystals with diamond structures composed of the alumina nanoparticles dispersed acrylic resins were fabricated as precursors through the stereolithographic additive manufacturing. The spatial resolution was approximately 0.5 %. Through the dewaxing and sintering processes, ceramic diamond structures were successfully obtained. The sintered sample is shown in Fig. 2. The deformation and cracking were not observed. The lattice constant of the diamond structure is 375 μm. The linear shrinkage was 24 %. It is possible to obtain the uniform shrinkage by designing an appropriate elongated structure in the vertical direction for compensation to the gravity effect. The relative density reached 97.5 %. The measured band gap frequencies were compared with calculation results by the plane wave expansion method. The measured frequency ranges of opaque regions corresponded to the calculation. According to the photonic band diagram, it was demonstrated that a complete photonic band gap opened between 0.40 THz and 0.47 THz. The modified diamond structure was fabricated by the three dimensional printing as shown in Fig. 3. The two and three period lattice structures were arranged on the right and left side of a plane defect, respectively. The terahertz wave was transmitted from the left side to right direction. In the measured transmission spectrum, one localized mode peak was observed in the band gap at the frequencies of 0.42 THz. The measured band gap region and peak frequency of the localized mode were compared with calculations by TLM method. They were in good agreement. The electric field distribution in the twinned diamond lattices was simulated. Figure 4 shows cross sectional images of the distributions. In this image, the red area indicates that the electric field intensity is high, whereas blue and green area indicates it is low. Incident terahertz wave is resonate and localized in the plane defect region between twinned diamond lattices. The amplified electromagnetic wave by multiple reflections can transmit through the photonic crystal. Therefore, the transmission peak will be formed in the band gap. On the right side of the sample, the radiation pattern shows the plane wave expansion. The micro photonic crystal with the twinned ceramic lattice of the diamond structure can be applied to a terahertz beam emitters.

The acrylic micro lattices with YSZ particles dispersion could be fabricated by the stereolithography as shown in Fig. 5. The formed precursors with dendrite structures were heated at various temperatures from 100 to 600 °C while the heating rate was 1 0 °C /min. The dewaxing process was observed in respect to the weight and color changes. The sintered solid electrolyte dendrite with the YSZ micro lattice structure is shown in Fig. 6. The deformation and cracking are not observed. The volume fraction of the air gaps is 50 % by the open paths. When the porous electrodes with random porous structure were sintered from the YSZ surly including polystyrene particles as foaming materials, it will be difficult to realize the prefect opened pores structures with the higher porosity over 40 % in volume fraction. In the dense microstructure of the YSZ lattice, the average grain size was approximately 4 μm. The relative density reached at 95 %. Micrometer sized cracks or pores were not observed. The obtained dense YSZ lattice structure will be exhibit the higher performances in mechanical properties as the porous electrodes of the solid electrolyte dendrites. The fluid flow velocities are visualized by using the FVM method as shown in Fig. 7. All air paths are opened for outsides and connected with each other in the YSZ dendrite lattice structures. The fluid flows can transmit toward specific direction smoothly. The pressure stress distributions in the dendrite were visualized. The fluid pressures are gradually distributed for flow direction, and the localization of the stress is not observed. The fabricated solid electrolyte dendrites with YSZ lattices can be considered to have higher performances as novel ceramic electrodes.

The acrylic resin structures incorporating the β-TCP particles were fabricated precisely following the designed model as shown in Fig. 8. Artificial bone scaffolds with dendritic lattice structures were designed, formed evaluated systematically to encourage osteoconductivity and tissue regeneration successfully[7]. Prosthetics mimic bones with ordered porous structures considering biological fluid flow behaviors could be processed. The micro ceramics rods are arranged periodically to create lattice patterns with various coordination number, and these aspect ratios should be modulated to realize biomimetic structures. The part accuracy of the lattices was measured under 50 μm in size difference. These were caused during the resin polymerization at the micro pattern exposures. The particles were homogeneously dispersed in the acrylic resin matrix. The dense ceramic structure was sintered successfully as shown in Fig. 9. The relative density of the sintered lattice was measured at 98 %. The linear shrinkage along the horizontal and vertical axes was 23 and 25 %, respectively. Fluid circulations in the perfect interconnected pores are simulated and visualized by fluid dynamic solver as shown in Fig. 10. Flow velocities in the spatial grids in the scaffold models are calculated through the FEM. The relation between the fluid velocity and cell deformation has been reported, and the high fluid velocity area in the scaffold is subjected to a shear stress which can assume the cell attachment difficulty. An isotropic distribution of flow rates are observed in the formed scaffold. And, this structure is insusceptible to a shear stress. Homogeneous tissue regeneration will be promoted through geometrical modification of the scaffold models. Hence, this is invaluable simulation for the tissue engineering before confirmative osteoconduction in vivo.

CONCLUSIONS

By stereolithographic additive manufacturing, micrometer order photonic crystals with a diamond structure composed of alumina dispersed acrylic resin were fabricated. The sintered alumina photonic crystal formed a complete band gap at the terahertz region. Localized modes were obtained by introducing a plane defect between twinned diamond structures. Solid electrolyte dendrites with yttria stabilized zirconia lattices were fabricated for anode electrodes of solid oxide fuel cells. Acryl precursors including ceramic particles were dewaxed and sintered through carefully optimized parameters. Solid electrolyte dendrites with opened air path networks exhibited effective transmission properties of fluid flows. Hydroxyapatite biological scaffolds with geometrical variations were fabricated successfully. The sintering body of scaffolds could precise reproduce exactly ordered porosities. Flow simulations will be powerful tools before time consuming clinical test. Structural modifications may help in realizing effective tissue regeneration in implanted tissue engineering. The ceramics additive manufacturing will be applied for various fabrications of functional structures and components in the near future.

REFERENCES

[1]K. Ohtaka, Energy Band of Photons and Low-energy Photon Diffraction, Physical Review B, 19 [10] (1979) 5057-5067.

[2]E. Yablonvitch, Inhibited Spontaneous Emission in Solid-state Physics and Electronics, Physical Review Letters, 58 [20] (1987) 2059-2062.

[3]S. John, Strong Localization of Photons in Certain Disordered Dielectric Superlattices, Physical Review Letters, 58 [23] (1987) 2486-2489.

[4]S. Kirihara, "Terahertz Wave Control Using Ceramic Photonic Crystals with Diamond Structure Including Plane Defects Fabricated by Micro Stereolithography", International Journal of Applied Ceramic Technology, Vol. 6, pp. 41-44, 2009

[5]Angel, S. M.; Kulp, T.J.; Vess, T. M. Remote-Raman Spectroscopy at Intermediate Ranges Using Low-Power cw Lasers, Appl. Spectrosc. 1992, 46, 1085-1091.

[6]Wang, H.; Wang, Q. Spectrum characteristics of nitrofen by terahertz time-domai spectroscopy, Phys. Conf. Ser. 2011, 276, 012209-012214.

[7]Kutteruf, M. R.; Brown, C. M.; Iwaki, L. K.; Campbell, M. B.; Korter, T. M.; Heilwail, E. J. Terahertz spectroscopy of short-chain polypeptides, Chem. Phys. Lett.2003, 375, 337-43.

[8]N. Q. Minh, "Ceramic Fuel Cells", Journal of American Ceramic Society, Vol. 76, No. 3, pp. 563–588, March, 1993.

[9]S. C. Singhal, "Solid oxide fuel cells for stationary, mobile, and military applications", Solid State Ionics, vol. 152-153, pp. 405-410, December, 2002.

[10]A. B. Stambouli, E. Traversa, "Solid Oxide Fuel Cells: a Review of an Environmentally Clean and Efficient Source of Energy", Renewable Sustainable Energy Reviews, vol. 6, No. 5, pp. 433–455, October, 2002.

[11]J. Hua, Z. Lüa, K. Chena, X. Huanga, N. Aia, X. Dub, C. Fub, J. Wanga, W. Su, "Effect of composite pore-former on the fabrication and performance of anode-supported membranes for SOFCs", Journal of Membrane Science, Vol. 318, No. 1-2, pp. 445-451, June, 2008.

[12]J. J. Haslam, A. Q. Pham, B. W. Chung, J. F. Dicarlo, R. S. Glass, "Effects of the use of pore former on performance of an anode supported solid oxide fuel cell", Journal of American Ceramic Society, Vol. 88, No. 5, pp. 513-518, March, 2005.

[13]T. Talebi, M. H. Sarrafi, M. Haji, B. Raissi, A. Maghsoudipour, "Investigation on microstructures of NiO-YSZ composite and Ni-YSZ cermet for SOFCs", International Journal of Hydrogen Energy, Vol. 35, No. 17, pp. 9440-9447, September, 2010.

[14]N. Komori, S. Tasaki, S. Kirihara, Fabrication of Dendritic Electrodes for Solid Oxide Fuel Cells by Using Micro Stereolithography, Proceedings of the 36th International Conference and Exposition on Advanced Ceramics and Composites, 33 (2012) 0813-1-6.

[15]C. Maeda, S. Tasaki, S. Kirihara, Accurate Fabrication of Hydroxyapatite Bone Models with Porous Scaffold Structures by Using Stereolithography, Materials Science and Engineering, 18[7] (2011) pp. 072017-1-4.

[16]S. Itoh, S. Nakamura, M. Nakamura, K. Shinomiya, K. Yamashita, Biomaterials 27 (2006), 5572-5579

[17]I. Sopyan, M. Mel, S. Ramesh, and K. A. Khalid, Science and Technology of Advanced Materials 8 (2007), 116-123

[18]A. L. Olivares, E. Marsal, J. A. Planell, and D. Lacroix, Finite element study of scaffold architecture design and culture conditions for tissue engineering, Biomaterials 30 (2009), 6142–6149

[19]A. J. F. Stops, K. B. Heraty, M. Browne, F. J. O'Brien, and P. E. McHugh, A prediction of cell differentiation and proliferation within a collagen–glycosaminoglycan scaffold subjected to mechanical strain and persuasive fluid flow, Journal of Biomechanics 43 (2010), 618-626

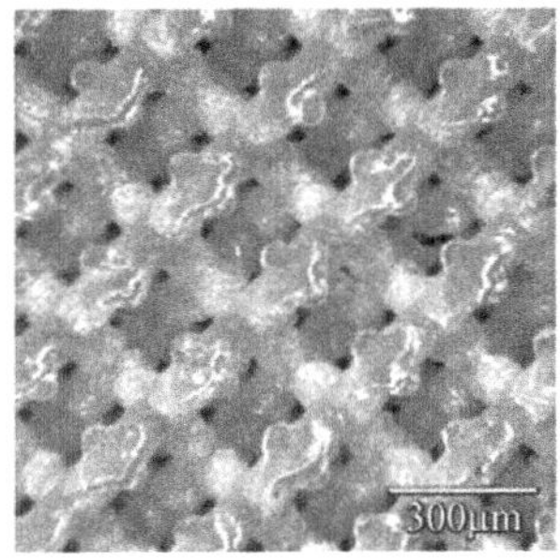

Fig. 2 An alumina model of the photonic crystal with a diamond structure fabricated using the micro stereolithographic additive manufacturing and nanoparticles sintering.

Fig. 3 Twinned photonic crystals composed of the sintered alumina lattices. The defect interfaces were formed between the diamond lattice domains parallel to the (100).

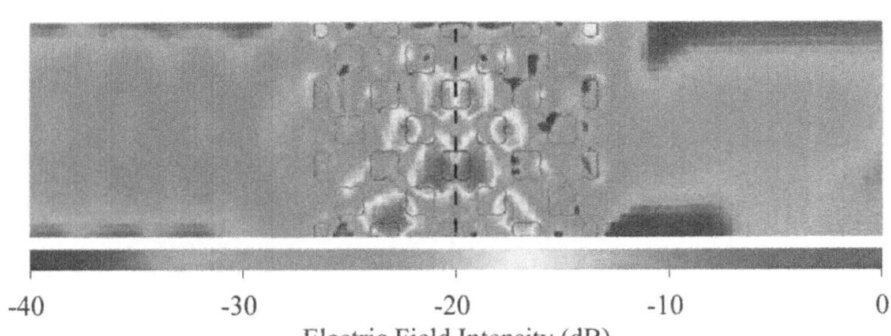

Fig. 4 Intensity profiles of electric fields in the twinned photonic crystal. The strong localized modes were formed in the defect interfaces through multiple reflections between the (100) plane.

Fig. 5 An acryl dendrite lattice of four coordination numbers including with YSZ particles dispersion fabricated by using the stereolithographic additive manufacturing.

Fig. 6 An YSZ dendrite lattice of solid electrolyte with a ordered porous structure. The precursor model was dewaxed and sintered with 2 μm in dimensional accuracy.

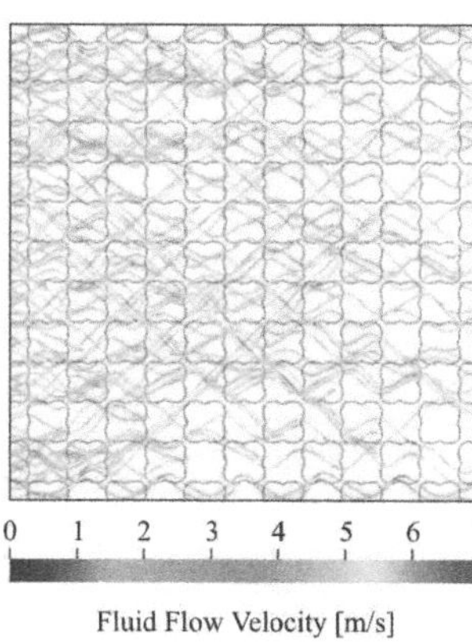

Fig. 7 A distribution of fluid flow velocities in the dendrite lattice structure simulated by FVM. The curved lines show the fluid flow paths according to the velocity vectors.

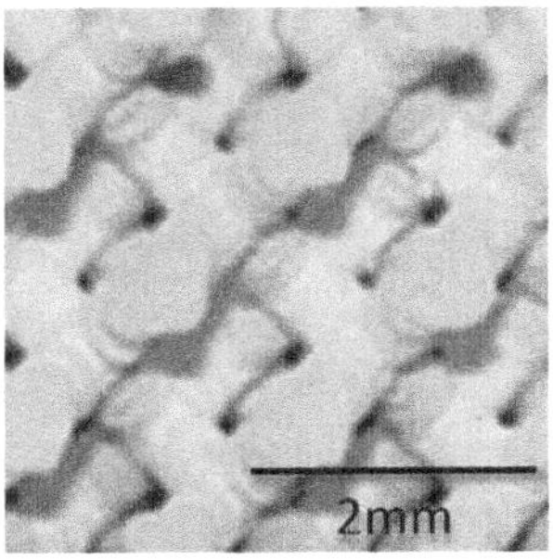

Fig. 8 An acryl lattice including with β-TCP particles formed by using the laser scanning stereolithography of additive manufacturing process.

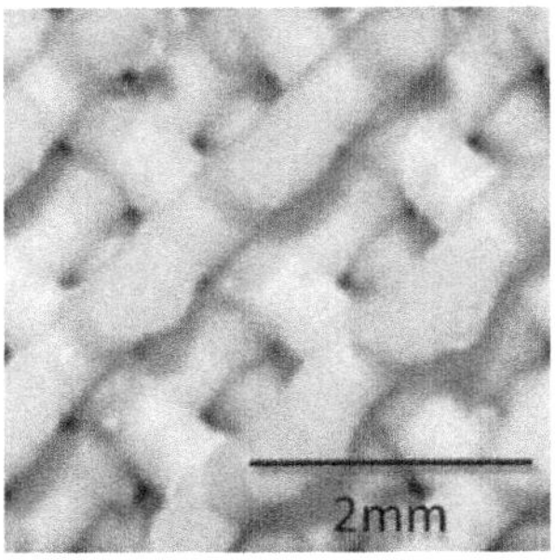

Fig. 9 An artificial bone with a β-TCP biological scaffold obtained through the dewaxing and sintering from the composite precursor.

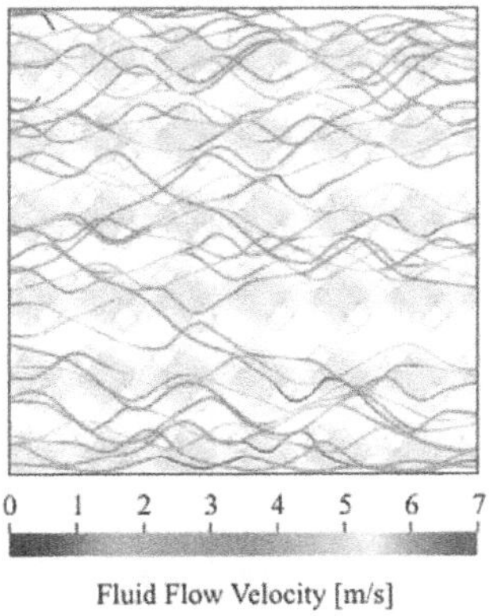

Fig. 10 Fluid flow behaviors of stream lines in the dendrite scaffolds with four coordination numbers simulated and visualized using the finite element method.

NEW LIGHTWEIGHT KILN FURNITURE - PRODUCTION PROCESSES AND PROPERTIES

U. Scheithauer, T. Slawik, E. Schwarzer, F. Tscharntke, H.-J. Richter, T. Moritz, and A. Michaelis

Fraunhofer Institute for Ceramic Technology and Systems IKTS Dresden, Germany

ABSTRACT

The main requirement on kiln furniture is the save support or transport of ceramic products during the thermal processing. Therefore, high compressive strength and high bending strength especially at high temperatures are needed. Thick dense or thick porous ceramic plates are state of the art. However, these structures have a high thermal mass, which has to be heated and quenched during each heat treatment cycle. Furthermore, a homogeneous temperature distribution in the kiln can only be realize at low heating or cooling rates.

Due to these challenges, the manufacturing of a novel generation of kiln furniture with advanced properties (low density and heat capacity, high stiffness, good thermal shock behavior) has been investigated. This new approach was realized by the combination of different green ceramic pre-products made by tape casting or extrusion. Planar green tapes were laminated with a supporting structure including macroscopic cavities to allow passage of air or gas and additionally reduce the density of the structure.

Due to the reduced heat capacity, the saving of energy and time up to 40% is possible, verified in different tests.

INTRODUCTION

Kiln furniture has to ensure the save support or transport of ceramic products during the thermal processing. Therefore, high compressive strength and high bending strength were needed, especially at high temperatures. Thick dense or porous ceramic plates are state of the art. However, a high thermal mass of these structures has to be heated. A homogeneous temperature distribution can only be achieved at low heating or cooling rates. Higher rates result in thermal induced mechanical stresses, which can damage the kiln furniture and the ceramic products.

To increase the possible heating and cooling rates and additionally to decrease the energy consumption to heat up the kiln furniture, a new generation of these products is needed. Several papers show alternative structures with macroscopic pores and open cavities allowing passage of gas or air through it.

Berroth presented structures made by extrusion with channels in one direction[1]. However, the realizable geometry (minimum thickness and constant cross-section) is limited and the structures are permeably only in one direction.

Lightweight structures made by a process known from paper technology are described by Travitzky et al.[2]. Thereby similar to the production of paper, cellulose fibres are used to produce a sheet while ceramic powder is used as filler material with high volume content. Due to the high content of ceramic powder, these tapes can be sintered like conventional ceramic green tapes with the advantage of forming same way like papers and realization of structures like corrugated boards. Lightweight structures can be created by laminating planar sheets on top of the corrugated structure. The disadvantage is the unavoidable burn out of the cellulose fibres leaving a porosity of 20-30 percent in the sintered structures[3], which lead to a significant decrease of strength. Known from lightweight applications (e. g. aviation) the used material should have a high stiffness and strength which cannot be guaranteed with this high porosity.

The purpose of this work was to develop new lightweight structures for kiln furniture with following properties:

- low thermal mass and heat capacity
- open macroscopic cavities allowing passage of air through it
- high stiffness and strength at elevated temperatures
- high thermal shock resistance

To achieve structures with the aspired properties tape casting and extrusion has been used for the production of green structures, which can be combined in the green state to sandwich-like structures with planar top and bottom layers and a supporting structure in between. After thermal processing, monolithic structures are available with a low heat capacity and a low density but a high stiffness.

USED SHAPING TECHNOLOGIES

Ceramic tape casting and honeycomb shaping

Ceramic tape technology was established for fabrication of large-scale thin substrates[4, 5]. Considering that, a homogeneous and de-agglomerated casting suspension is necessary. An alumina suspension was prepared and cast to green tapes. Haderk et al.[6] described the composition and production of the suspension and tapes in detail. The produced planar tapes can be converted by bending, cutting or embossing and laminated to form 3D-structures[7].

Honeycomb structures are known as lightweight structures characterized by a very high stiffness-to-mass-ratio. Nevertheless, the passage of air through the separated cells is not possible after adding top and bottom layers to complete the sandwich structure.

Alternative structures, which have a similar design to honeycombs but also allows the passage of gas and can be produced using ceramic green tapes, were developed. A support structure made by punching and sticking of green tape stripes with gaps to form an open grid-like structure was describe in a previous paper[7].

With the aim of continuous production of honeycomb structures, a special technology known from paper industry was adapted. This technology is usually uses for the production of paper-based honeycomb structures, which are applied as stiff but lightweight wood furniture (tables or doors) or packaging materials.

The adapted shaping technology was described in a former work[8]. Different single tapes, partially laminated, form a stack, were cut into stack stripes and perforated by punching (Figure 1). After lamination of a number of stack stripes a nearly endless stack results which can be stretched to realize the honeycomb structure. In a final step, top layer and bottom layer are laminated (Figure 2 and 3) and the green component is completed. Finally, this structure is debinded and sintered.

Figure 1: schema of first steps to shape the honeycomb structure from green tapes – partial lamination, stacking, and punching

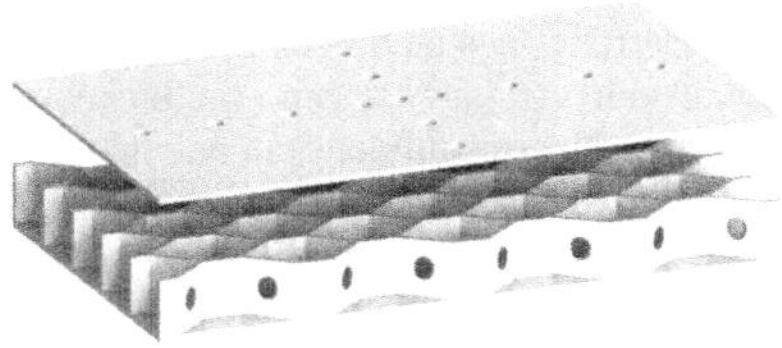

Figure 2: laminating of top and bottom layer on stretched honeycomb structure, schema

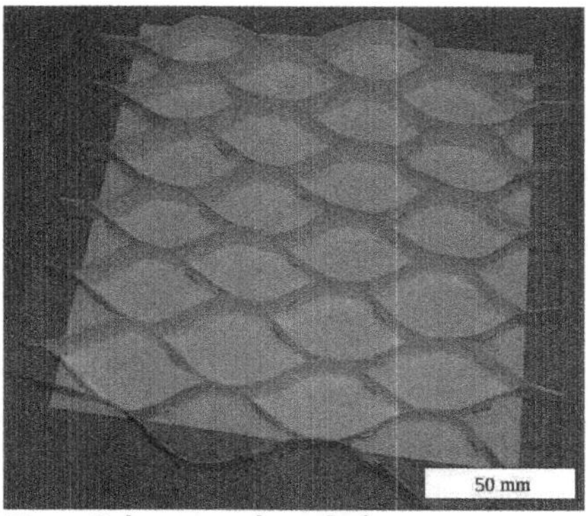

Figure 3: honeycomb made from green tapes as supporting structure laminated on planar multilayer as bottom layer of the kiln furniture

Figure 4: scheme for kiln furniture with tapes as top and bottom layer and vertical pipes as supporting structure in between

Figure 5: scheme for kiln furniture with multilayers as top layer (1) and bottom layer (4), horizontal pipes as supporting structure in between (2), and joining suspension (3)

Viscous polymer processing

Viscous polymer processing (VPP) was used to produced pipes as other kind of supporting structure which can be arranged horizontally or vertically between the top and bottom layers (Figure 4 and Figure 5). This production technology is typically used to fabricate structural and functional ceramics with excellent mechanical properties and improved electrical properties[9, 10] like PZT-components[11].

Feedstocks for VPP can be processed e.g. by extrusion. Extruding is a well-known and established ceramic shaping method for many years[12]. Especially the production of tubular or other elongate parts with wide variety of cross-sections use the described method. For example, ceramic honeycombs for diesel particulate filters or catalyst carriers and multi-channels tubes are made by extrusion of cold-plastic ceramic feedstocks. The rheological properties of ceramic feedstocks strongly depend on production process (mixing, kneading) and on the feedstock composition. The feedstock properties and the extrusion parameter need to be adjust to the particular extruded profile to obtain defect-free parts also at high capacity of extrusion process.

The VPP technology uses water-solutable polymers like cellulose ether as binder system which can be processed at room temperature. Very high shear stress can be realized within the viscoelastic ceramic/polymer feedstock to deagglomerate the ceramic raw powders, but also strong extrusion pressures are necessary for the shaping of the ceramic green bodies[11].

MANUFACTURING OF NEW KILN FURNITURE

Tape casting

Ceramic green tapes (Figure 6) were produced from Al_2O_3-powder (Almatis CT3000SG, d_{50}=0.4 µm) by doctor blade method. The water-based suspension had a powder content of about 23 volume-%. The casting gap was 1200 µm and after drying the tapes had a thickness of about 450 µm. The suspension preperation and tape casting process are discribed by Haderk et al. in detail[6].

Cold-plastic extrusion

Feedstocks with about 45 volume-% Al_2O_3-powder (Almatis CT3000SG, d_{50}=0,4 µm) were prepared, extruded with a vacuum single-screw extruder and dried encapsulated and propped for three to five days at room temperature to avoid defects and warpage. The extruded pipes had an outer diamter of 10 mm and an inner diameter of 7 mm (Figure 7). After drying the powder content was about 68 volume-%, which corresponded to the powder content within the dried green tapes to avoid different shrinkage behaviour during sintering.

The dried extrudates were cut to length by using a bandsaw.

Figure 6: alumina green tape after drying

Figure 7: extruded alumina pipe (10 mm outer and 7 mm inner diameter)

Lamination of green tapes and joining of tapes and extrudates

The top and bottom layers of the kiln furniture were manufactured by lamination of two or more green tapes. These multilayers guarantee a higher stiffness and strength in the green state as well after sintering then a single tape with the same thickness as the multilayer.
To laminate the tapes a special joining suspension was used[13], which is filled with alumina particels, too. The same joining suspension was also used to connect the green extrudates and tapes.

For each lamination process, one green component was coated with the joining suspension completely or partially. Afterwards, the second green component was added as lamination partner and a hand roller or a pair of rolls applied a pressure to connect the lamination partner simultaneous removing the unnecessary joining suspension as well as the embedded air bubbles.

Figure 8 and Figure 9 show different green structures, which were manufactured from green tapes and extruded components.

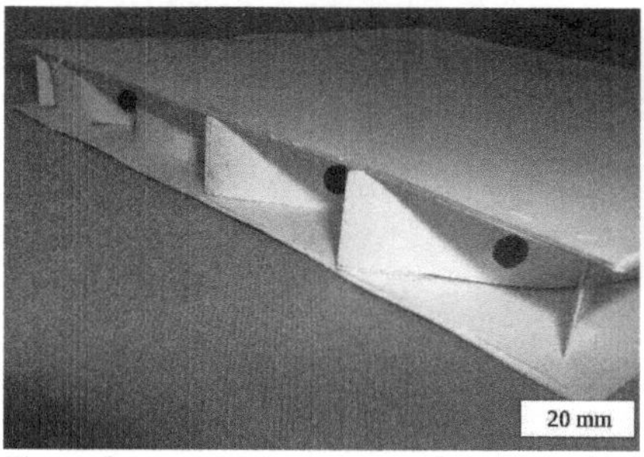

Figure 8: green component made by the combination of planar multilayer and vertical green tapes formed to honeycombs according to Figures 1 - 2

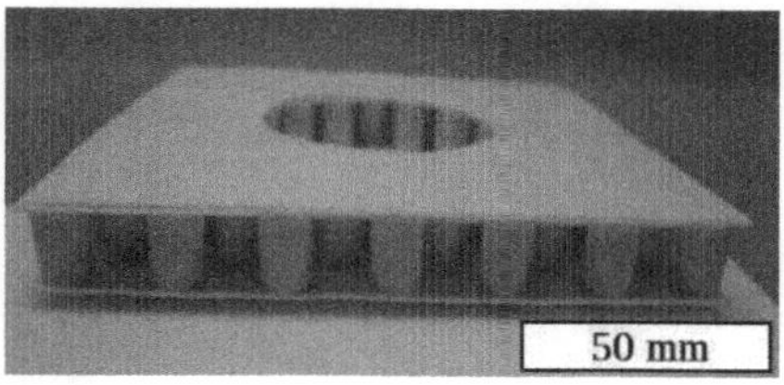

Figure 9: green component made by the combination of planar multilayer and vertical extrudates

Debinding and Sintering

The thermal debinding of all structures happened under air at temperatures up to 400°C , with heating rates between 0.2K/min and 0.5K/min. The sintering happened also under air at a temperature of 1700°C or 1750°C for 1 h with a heating rate of 1K/min and a cooling rate of 3K/min.

CHARACTERIZATION OF NEW KILN FURNITURE

The figures 10 to 12 show different types of the new kiln furniture after sintering. No cracks are visible and delaminations between the support structure and the top and bottom layers could not be observed. With a higher number of tapes used for the multilayers, the stiffness of these multilayers increases in the green and sintered state resulting in a better surface quality concerning planarity. Furthermore, the higher thickness of the top and bottom layers allows the grinding of these surfaces to realize a planar surface with a very high planarity (Figure 13).

Holes within the top and bottom layers, which inserted in the green tapes or multilayers, allow convection of the kiln atmosphere through the kiln furniture, which results in a better temperature distribution (see below) and a better removal of debinding products.

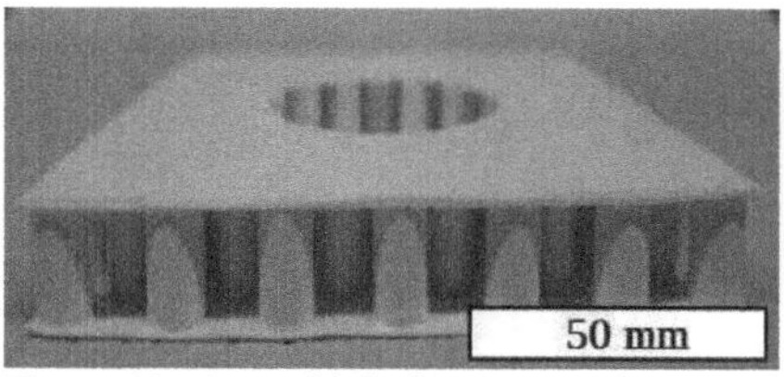

Figure 10: sintered kiln furniture with vertical pipes as supporting structure, as fired

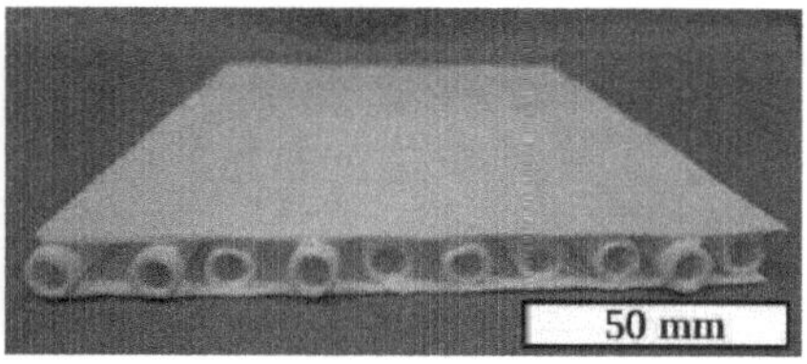

Figure 11: sintered kiln furniture with lying pipes as supporting structure, as fired

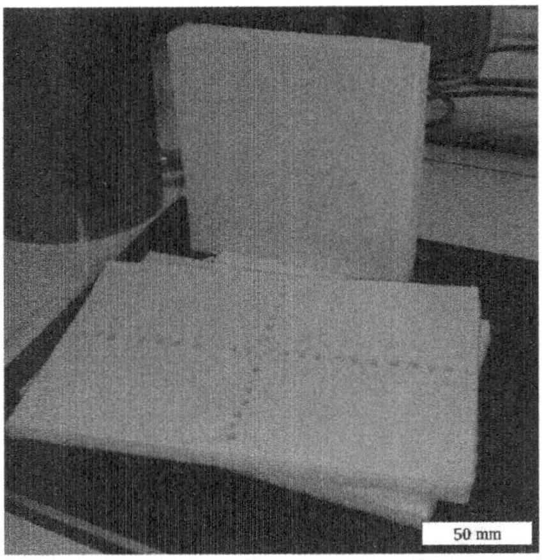

Figure 12: sintered kiln furniture with honeycombs as supporting structure, as fired

Figure 13: sintered kiln furniture with vertical pipes as supporting structure, grinded

Different tests happened in a laboratory kiln to evaluate the possible saving on energy and time by the usage of the new structures in comparison with dense alumina plates. The stiffness of the new structures was characterised by creep tests, too. This creep test was described in a former paper[7] and showed that no significant deformation of the kiln furniture were measured after applying a load which was two- and four-times higher than the mass of the kiln furniture. Temperatures of 1400°C were hold for 50h each.

To evaluate the possible saving of energy and the influence of the kiln furniture on the temperature distribution within the kiln some tests in a laboratory kiln happened. The energy consumption and the real temperature in the centre of the kiln were monitored for three different charges. First, there was an empty drive, second a dense plate placed within the kiln to represent the state of the art (Figure 14) and third a new structure with tapes as supporting structure was used (Figure 15). For all tests, the same temperature regime with heating and cooling rate of 3K/min and two dwell times for 1 h each at 400°C and 1000°C was used. The test setup was similar to the creep test, but no load was applied on the upper kiln furniture.

The thermocouple of the kiln is fixed in the left upper part of the kiln chamber and the thermocouple used for monitoring the real temperature was placed in the centre between the two kiln furniture laying on the lower kiln furniture.

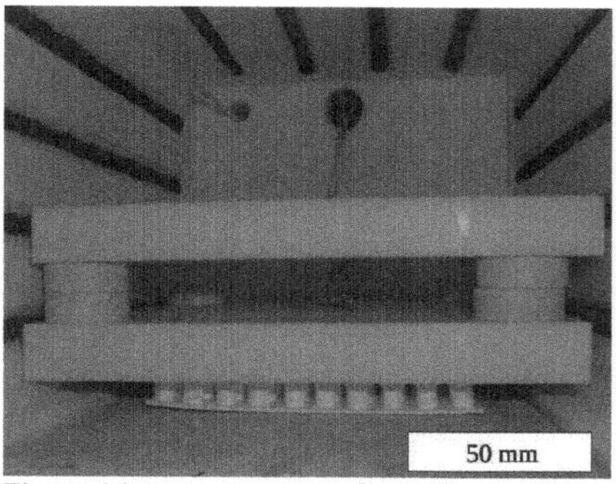

Figure 14: test setup to evaluate the possible energy saving, dense alumina plates

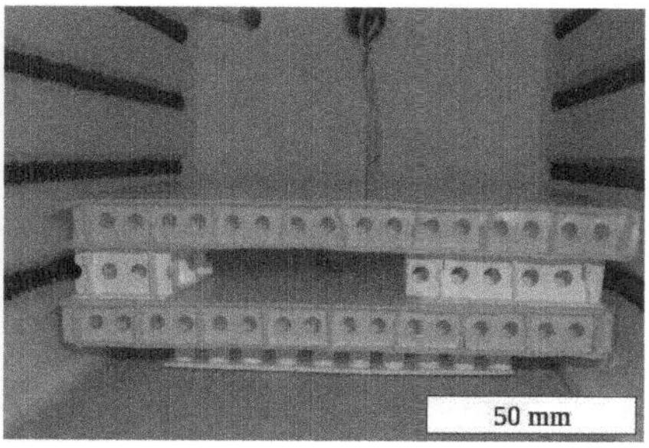

Figure 15: test setup to evaluate the possible energy saving, new kiln furniture

Figure 16 shows the monitored temperature for different tests. Using the dense alumina plate the real temperature differed very much from the set temperature during the heating and cooling process, especially at lower temperatures, and it takes a long time to achieve target temperature levels. In comparison to this, the curve for testing the new kiln furniture is nearly the same as the curve for an empty drive of the kiln. The time saving for the cooling from 1000°C to 100°C was about 40%.

Figure 17 shows the relative energy consumption for the three tests. The energy consumption for the dense plate is set to 100 and plot relative to this for the two others. If the new kiln furniture is used the energy consumption for heating is significant lower (- 40%) than using dense plates and nearly the same as for the empty drive.

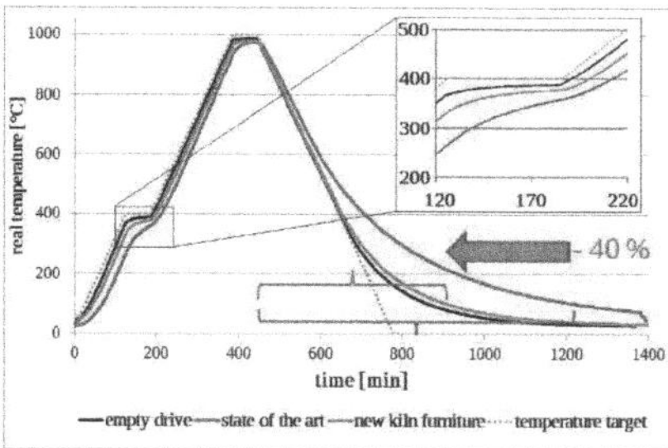

Figure 16: Temperature distribution within the kiln for dense and new kiln furniture

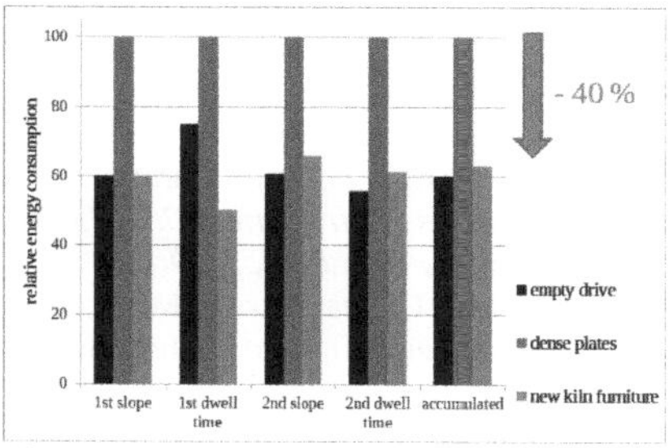

Figure 17: Energy consumption for dense and new kiln furniture

SUMMARY

The paper shows that it is possible to produce novel lightweight components with a sufficient stiffness, used as kiln furniture with a significant lower mass compared to the state of the art products. These components allow higher heating and cooling rates due to a thin and open structure. Tests in a laboratory kiln showed the possibility for significant saving of energy and time (-40%) during thermal treatment by using the new kiln furniture. Al_2O_3-demonstrators were manufactured with a size of 225x140x15mm^3 and a mass of about 230g. This corresponds to an overall density of about 0.5g/cm^3.

New production technologies allows a reasonable production of the kiln furniture. One technology was adapted from the paper industry and transferred to the processing of ceramic green tapes. The other bases on the combination of extrusion and tape casting as two very productive shaping technologies.

The next step will be to produce the new kiln furniture in quantity by using the new technologies and to quantify the possible energy savings, the possible heating and cooling rates and the stiffness of the structures under real conditions. Furthermore, the suitability of these technologies for the production of non-planar kiln furniture with a low density should be investigated.

ACKNOWLEDGMENTS

This project is kindly supported by the German Research Foundation (DFG) under the reference number MI 509/10-2 within the Priority Program SPP 1418 "FIRE" and by the Federal Ministry of Education and Research of Germany within the "ExtruFol"-project (FKZ 01LY1211A).

REFERENCES

1. Berroth, K. (2005). Komplexe Strukturen aus Hochleistungskeramik. *cfi/ber. DKG 82*
2. Travitzky, N., Windsheimer, H., Fey, T., & Greil, P. (2008). Preceramic Paper-Derived Ceramics. *J. Am. Ceram. Soc., 91 [11],* 3477–3492.
3. Gutbrod, B., Haas, D., Travitzky, N., & Greil, P. (2011). Preceramic Paper Derived Alumina/Zirconia Ceramics. *Advanced Engineering Materials, 13, No. 6,* 494-501.
4. Hotza, D., & Greil, P. (1995). Review: Aqueous tape casting of ceramic powders. *Material Science and Engineering A202,* 206 – 217.
5. Mistler, R. E., & Twiname, E. R. (2000). Tape casting Theory and Practice. *The American Ceram. Society,* ISBN: 1-57498-029-7
6. Haderk, K., Scheithauer, U., Richter, H.-J., Petasch, U., & Michaelis, A. (2012). Development of Ceramic Tapes for Thermal Shock Resistant Calcium Aluminate Refractory Materials with Graded Porosity. *Refractories Manual 2/2012,* 84 – 8.7
7. Scheithauer, U., Freytag, C., Haderk, K., Moritz, T., Zins, M., & Michaelis, A. (2014). Novel Generation of Kiln Furniture. *Proceedings of the Unified International Technical Conference on Refractories (UNITECR 2013), Energy Savings through Refractory Design, Edited by Dana G. Goski, Jeffrey D. Smith, Chapter 63,* 361-365
8. Scheithauer, U.; Dannowski, M., Schwarzer, E., & Slawik, T. (2014). Innovative Kiln Furniture, their Influence on the Temperature Distribution within the Kiln, and a New Production Technology. *Refractories Manual, Special Technologies,* 312-316
9. Alford, N. McN., Birchall, J.D., & Kendall, K. (1987). High-strength ceramics through colloidal control to remove defects. *Nature, 330,* 51–53.
10. Su, B., Ponton, C.B., & Button, T.W. (2001). Hydrothermal and elctrophoretic deposition of lead zirconate titanate (PZT) films. *J. Eur. Ceram. Soc., 21,* 1539–1542.
11. Su, B., & Button, T.W. (2009). A Comparative Study of Viscose Polymer Processed Ceramic Based on Aqueous and Non-Aqueous Binder System. *J. Mater. Process. Technol., 209 [1,]* 153–157.
12. Händle, F. (2009). Extrusion in Ceramics. *Springer, Berlin*
13. Scheithauer, U., Haderk, K., Richter, H.-J., Petasch, U., & Michaelis, A. (2011). Influence of the kind and amount of pore forming agents on the thermal shock behaviour of carbon-free refractory components produced by multilayer technology, *refractories WORLDFORUM, Volume 4, Issue 1,* 130-136.

THE ROLE OF CALPHAD APPROACH IN THE SINTERING OF B_4C WITH SIC AS A SINTERING AID BY SPARK PLASMA SINTERING TECHNIQUE

Mohammad Asadikiya, Christopher Rudolf, Cheng Zhang, Benjamin Boesl, Yu Zhong

Department of Mechanical and materials engineering, Florida International University, USA

ABSTRACT

Dense sintered B_4C is an attractive material for a variety of critical applications in different industries. Using sintering aids is a typical approach to help increase the relative density. SiC is one of the common additives for this purpose. In this study, spark plasma sintering (SPS) technique was used to sinter B_4C+9wt%SiC mixture powder at 1750°C under the mechanical load of 60 MPa with 20 min holding time and 100°C/min heating rate. A pure B_4C control sample was also sintered with the same parameters for comparison. The phase stability and changes were discussed with the CALPHAD approach and the behavior of phases in the system during sintering was discussed. It was indicated that a general understanding about the behavior of phases around sintering temperature and also the amount and effectiveness of additives can be extracted by applying computational thermodynamics. It was also discussed that although there is a sintering temperature difference between the SPS and pressureless techniques, the CALPHAD approach can provide reasonable information for applying in the SPS.

INTRODUCTION

Boron carbide (B_4C) has a lot of industrial applications because of its excellent properties such as high melting point, high hardness, low density and chemical stability. B_4C is the hardest material after diamond and cubic boron nitride (c-BN). B_4C has various applications including lightweight armors, abrasive powder, blasting nozzles and grinding wheels[1-3]. However, due to covalently bonded structure of B_4C, it is difficult to sinter this material[4]. Thus, reaching full theoretical density and having the maximum hardness is a challenging issue for this material.

During recent years, different groups have tried to increase the relative density of sintered B_4C. One of the methods they used is adding certain amount of sintering aids in B_4C. The additives like Si, SiC and Al_2O_3 can reduce the sintering temperature and increase the density of the product[5-7]. Furthermore, a relatively novel sintering method i.e. Spark Plasma Sintering (SPS) has been used in which the sample could be sintered by passing a pulsed direct current through the sample. Indeed, the temperature increases based on Joule heating and a sintered sample is produced by assistance of high pressure which applies to the powder spontaneously[8]. Researchers have tried to adjust the SPS parameters like heating rate, maximum temperature and holding time to reach highest density and as a result, better mechanical properties.

The type of additive and amount of it are of great importance and a typical method which people usually use to find out them is try and error approach. This method, however, is time consuming and costly. In this study, the CALPHAD (Calculation of Phase Diagrams) approach was used to analyze the effectiveness of a sintering aid and appropriate amount to be applied. The effect of SiC as an additive for B_4C sintering was investigated in this study.

EXPERIMENTAL PROCEDURE

Boron carbide powder (Grade HD 15, D50 particle size: 0.6-0.9 μm, H. C. Starck, Germany) was used in the present study as received. The additive used in this study was SiC (LAP5-0002-010-2600, Lapmaster Wolters, USA). The additive was added to B_4C in the amount of 9 wt%. The mixture was then ball milled in a planetary ball milling machine (Across International, VQ-N, USA) for 3 hours at a rotary speed of 1200 rmp. The mixture was poured into a graphite die with inner diameter of 20 mm. The starting pressure loaded on the powder was 5 MPa and after that, gradually increased to 60 MPa. The designated maximum temperature was 1750˚C, the holding time at this temperature was 20 min and the heating rate was 100˚C/min. In addition to the sample with additive, a sample was sintered without adding any additive with the same parameters mentioned above labeled sample A. The sample with the SiC was labeled sample B. The density of samples were measured by the Archimedes principle. The microstructure of the fracture surface was investigated by using scanning electron microscopy (SEM, JSM 6330F, JEOL, Japan). The hardness of the samples was also measured by micro hardness machine under 1000 grams of load applied for 15 seconds (Wilson Tukon 200).

THERMODYNAMIC MODELING

The CALPHAD approach relies on modeling the Gibbs energies of each of the individual phases in the system. This characteristic state function is of particular interest because under constant temperature and pressure, the Gibbs energy is minimized at equilibrium, and temperature and pressure are the variables that are typically controlled experimentally. The thermodynamic databases based on the Gibbs energies are then constructed using experimental data and software programs like Thermo-Calc[9].

Boris Kasper in 1996 assessed the B-C-N-Si thermodynamic database[10] and we applied B-C-Si part from that database in this study. This database was applied by Thermo-Calc software version 4.1[11] to analyze the effect of SiC as a sintering additive on B_4C sintering. In this study, B-C-Si isothermal ternary phase diagrams were calculated in different temperatures to see the phase changing in the interested areas. The B_4C-SiC pseudo binary phase diagrams were also calculated to see the phases on B_4C-SiC line at each temperature. Based on the calculated diagrams, the effect of SiC addition on the sintering of B_4C was discussed.

RESULTS AND DISCUSSION

Figure 1 shows the Ram displacement, pressure and temperature, all vs. time during the SPS process for sample B (with SiC).

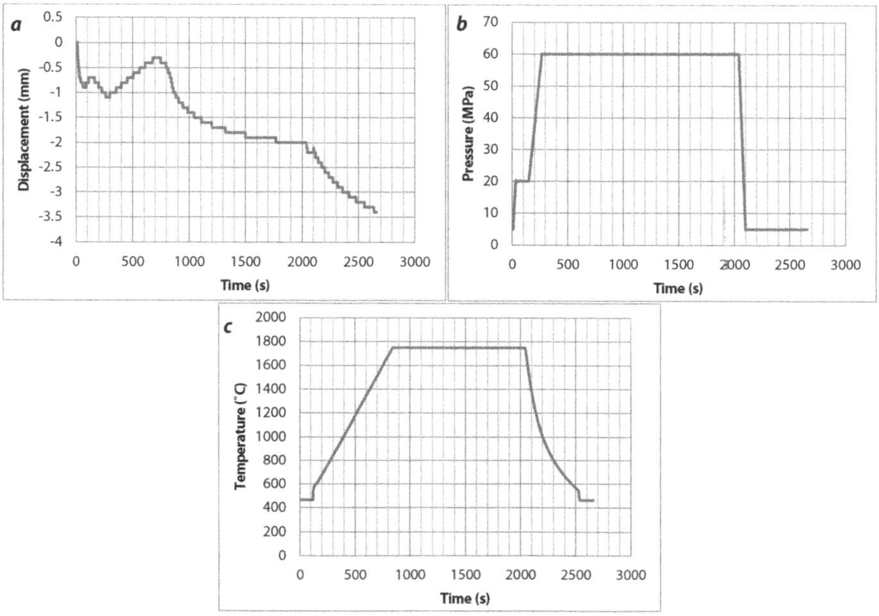

Figure 1. SPS parameters changing for sample B. a) Ram Displacement, b) Pressure, c) Temperature all vs. time.

The behavior of the parameters for sample B during SPS is described as below. The first data right after applying the 5 MPa load was considered as the initial point in all diagrams shown in Figure 3. By increasing the pressure vs. time, the ram starts to move and shrink the powder. Based on the software program of our SPS machine, it holds the pressure at 20 MPa until the temperature reaches 600°C and then starts to increase. A small moving up of the ram during holding on 20 MPa is related to thermal expansion. As the pressure increases, the temperature goes up to reach the maximum temperature i.e. 1750°C. Movement of the ram vs. time is shown in figure 1 (a). The movement of the ram stops once the pressure reaches at 60 MPa and then, it starts to move up since the temperature is still increasing and thermal expansion happens in both die and powder and also punches. It should be mentioned that the movement of the ram is the resultant effect of shrinkage and thermal expansion during the process.

Before reaching the maximum temperature which is 1750°C, the ram stops moving and starts to shrink the powder. Based on the diagram (a) in figure 1, the speed of shrinkage is fast at the beginning and becomes slow to reach a plateau at the end. The last ram movement before the end of holding time is at 1770th second from the start of process. As soon as the holding time is finished and sample starts to cool down, ram movement is observed which is related to thermal shrinkage in compensation of thermal expansion which happened during the process.

Table 1 shows the relative density and hardness of the samples with and without sintering aid. As it is shown in this table, addition of SiC to B_4C caused increase in density and hardness of the sample. Figure 2 indicates the fracture surface of both samples. By comparing the two micrographs, it is apparent that the picture related to B_4C+SiC (sample B) structure has higher density.

Table I: Relative density and hardness of the samples.

Sample code	Relative Density	HV (GPa)
A (without additive)	89.5%	15.42
B (with 9 wt% SiC)	95.5%	21.7

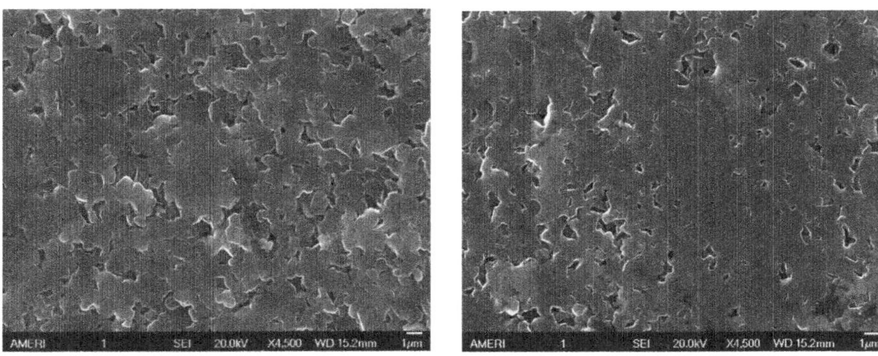

Figure 2. Fracture surface of samples A (left) and B (right)

Figure 3 shows three B-C-Si isothermal ternary phase diagrams calculated with Thermo-Calc at 2230, 2240, 2250 and 2260°C. Each of these diagrams shows the stable phases for each composition at a fixed temperature.

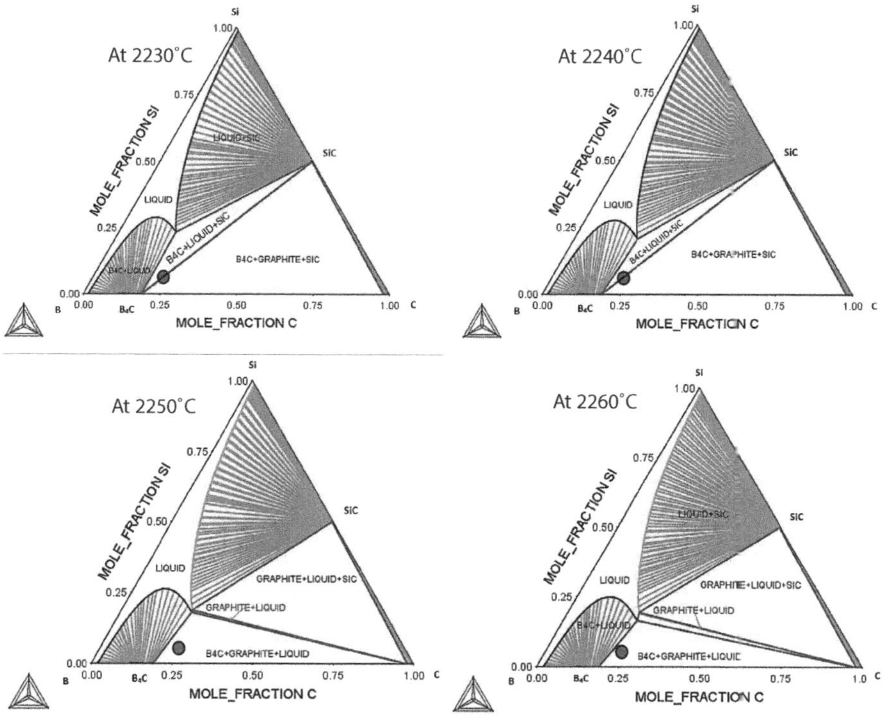

Figure 3. Isothermal B-C-Si phase diagrams at 2230, 2240, 2250 and 2260°C. The position of sample B is indicated by ● .

It is worth noting that the authors are fully aware of the differences of SPS conditions and diagrams in figure 3. The ternary diagrams in figure 3 were plotted at the pressure of 1 atmosphere, however, there is vacuum condition in SPS. In addition, the current density and Joule heating both have effect on the sintering[8] which are not considered in the diagrams. Moreover, in SPS, a mechanical load is applied to the powder. All these conditions cause the sintering temperature to be considerably reduced from around 2250°C in pressureless sintering to around 1800°C in SPS technique. Despite of these conditions, the phases involved in a system should reach a specific situations in order to enhancing diffusion, making neck at the particles' junction and as a result, sintering. Therefore, it is possible to correlate the phase behavior around sintering temperature in SPS process and plotted diagrams. Hence, a clear understanding about the phases of a system around sintering temperature can be provided using the CALPHAD approach since the behavior of phases around sintering temperature in SPS technique is almost the same as that of in the diagrams plotted in figure 3. As it is shown in figure 3, for the diagram at 2230°C, there is B_4C+SiC line with B_4C+Liquid+SiC three phase region on the left and B_4C+Graphite+SiC three phase region on the right. The amount of liquid in the three phase region on the left increases as the temperature goes up to reach 2240°C which is indicated in the related diagram in figure 3. The liquid region is extending as the temperature increases and theoretically a high amount of liquid at the temperatures below sintering temperature is not desirable. According to figure 3, when the temperature increases to 2250˚C, the system enters to B_4C+Graphite+Liquid region. This indicates that a small amount of liquid appears in the sample which is helpful to enhance sintering of the powder. Since SiC has lower melting point than B_4C, it partially melts around sintering temperature and provides small amount of liquid phase in the system.

The area related to sample B is highlighted in Figures 3 and 4. As it was mentioned above, the amount of added SiC is 9 wt% and this amount of SiC is equal to 0.119 mole fraction as indicated with the dash line in figure 4. Based on figure 4, sample B will enter to B_4C+Graphite+Liquid three phase region around 2250˚C (which is around sintering temperature in the conditions of this diagram). The amount of liquid in this region can be controlled by adjusting the amount of additive. According to figure 4, as the mole fraction of SiC increases toward the eutectic point on the eutectic line, the amount of liquid increases. Therefore, the amount of additive should be controlled to adjust the amount of liquid phase based on the phase diagram. A little amount of liquid which will be provided by adding a little amount of SiC into the B_4C is helpful to enhance sintering of B_4C.

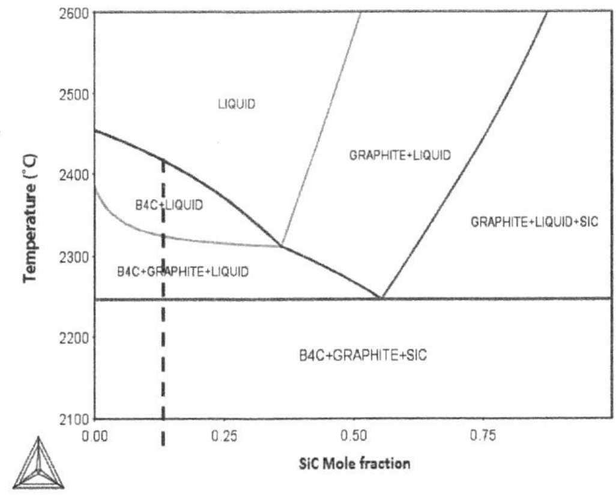

Figure 4. B_4C-SiC Pseudo binary phase diagram in 1 atm.

This liquid phase enhances densification since the diffusion becomes easier and also movement of particles will be enhanced. Liquid phase can easily fills the pores and void spaces between particles. All these effects can contribute to increase the density of the sample.

CONCLUSIONS

It was indicated that by using of CALPHAD approach, a better understanding of the additive effect for B_4C sintering can be obtained. It is possible to estimate an appropriate amount of sintering aid to be added in order to receive the maximum advantage of adding the additive. Being aware of the stable phases and their amounts at each temperature and in each mole fraction of the additive helps determine the amount of additive wisely by which the sintering and densification are enhanced. In addition, it was shown that even though the sintering temperature is much lower in SPS technique, the CALPHAD approach is helpful to provide better understanding of the behavior of phases around sintering temperature.

REFERENCES

[1]F. Thevenot, J. Euro. Ceram. Soc. 6[4] (1990) 205-225.
[2]F. Thevenot, Key Eng. Mater. 56-57 (1991) 59-88.
[3]M. Grujicic, B. Pandurangan, K. L. Koudela and B. A. Cheeseman, App. Surf. Sci. 253[2] (2006) 730-745.
[4]H. Lee, R.F. Speyer, Pressureless sintering of boron carbide, J. Am. Ceram. Soc. 86 (2003) 1468–1473.
[5]F. Ye, Z. Hou, H. Zhang and L. Liu, J. Am. Ceram. Soc. 93[10] (2010) 2956–2959.
[6]F. C. Sahin, B. Apak, I. Akin, H. E. Kanbur, D. H. Genckan, A. Turan, G. Goller and O. Yucel, Solid State Sci.
14 (2012) 1660-1663.
[7]Y. Çelik, G. Göller, O. Yücel and F. Şahin, Adv. Sci. Tech. 63 (2010) 79-84.
[8]B. M. Moshtaghioun, F. L. Cumbrera-Hernández, D. Gómez-García, S. de Bernardi-Martín, A. Domínguez-
Rodríguez, A. Monshi and M. H. Abbasi, J. Eur. Ceram. Soc. 33 (2013) 361–369.
[9]B. Jansson, Trita-Mac-0234, Royal Institute of Technology, Stockholm, 1984.
[10]B. Kaspar, Ph.D. Thesis, Universität Stuttgart, Stuttgart, 1996.
[11]B. Sundman, B. Jansson, J.-O. Andersson, CALPHAD 9[2] (1985) 153–190.

Powder Processing Technology for Advanced Ceramics

EFFECTIVE EXFOLIATION OF LAMINATED h–BN PARTICLES BY A NOVEL ROTATING DISK METHOD

Yuichi Tominaga, Daisuke Shimamoto, Kimiyasu Sato, Yusuke Imai, and Yuji Hotta
National Institute of Advanced Industrial Science and Technology (AIST),
Moriyama-ku, Nagoya 463-8560, Japan.

ABSTRACT

In order to investigate the exfoliation of laminated hexagonal boron nitride (h–BN) particles, a wet–rotating disk (WRD) method was used as a novel high shearing method. The sedimentation test of milled h–BN particles indicated that the sedimented cake heights of h–BN particles prepared by the WRD method was much higher compared to raw h–BN particles. Also, the particle size distribution of h–BN particles exfoliated by this method was relatively as similar as that of raw h–BN particles. Furthermore, h–BN particles treated by the WRD were observed by a field emission scanning electron microscopy (FE–SEM), resulting that the decrease of h–BN thickness was affected to the sedimented cake height in the sedimentation test. The thickness of WRD milling-assisted h–BN was thin compared to raw materials. The aspect ratio of exfoliated h–BN particles was increased more than 80 percentage that of raw h–BN particles. Moreover, the sedimented cake heights of prepared h–BN particles were affected to rotation speed, number of flow pass, and clearance between two rotation disks. These various results indicated that the WRD process led to exfoliation of the laminated h–BN particles effectively.

INTRODUCTION

Laminated ceramics particles, such as hexagonal boron nitride (h–BN), have been widely used as filers for polymer ceramics which are lightweight materials with high electrical and thermal conductive properties[1-8]. Moreover, in case of ceramics fields, h–BN particles have been utilized for composite ceramics with high performance of mechanical properties[9, 10]. Laminated ceramics particles are stacked in numerous 2D single layers by weak van der Waals interaction to form a 3D crystal[11]. In order to utilize the excellent properties of h–BN particles, laminated ceramics particles must be exfoliated to thin layers[12, 13]. The exfoliation of laminated ceramics particles causes high aspect ratio as nanosheets with high specific surface area, and leads to improvement of various performances of composite materials.

Since the original Scotch tape method has been reported for exfoliation of graphene by Novoselove's group[14], several exfoliation methods have been tried to exfoliate the laminated ceramics particles[15-19]. For example, Paciles's group applied the Scotch tape method to exfoliate h–BN particles[16]. In this method, h–BN nanosheets with the best combination of thickness and lateral size could be exfoliated by repeating the peeling–pressing process. Also, Li's group has synthesized the h–BN nanosheets by ball milling method in liquid phase[18]. Furthermore, Lin's group has exfoliated the h–BN by using sonication method[20]. However, as one of considerable problems, ball milling and sonication methods are process due to strong energy, causing that the crystal structure of h–BN particles may be broken. So, the exfoliated h–BN particles have a lot of defects and lead to the poor properties of composite materials[18, 19]. Furthermore, the mechanical shear force occurred by these methods is applied to the destruction of the basal phase not only the exfoliation, resulting that the aspect ratio of laminated ceramics particles is increased hardly. Therefore, the development of a novel method is really necessary to supply the exfoliated laminated ceramics particles without destroying the crystal structure of laminated ceramic particles.

Recently, a wet–rotating disk (WRD) milling has been utilized as a novel method of mixing, dispersion, and pulverization in chemical engineering and powder technology fields[21-23]. In the WRD method, the slurry or suspension sandwiched between the parallel disks is able to add a strong shear flow at high-speed rotation of the disk. Using the effect, the homogeneous dispersion and the pulverization of agglomerated spherical ceramic particles in slurry are achieved within a short time.

In this study, in order to investigate the exfoliation of laminated ceramics particles by the WRD milling, we demonstrated to exfoliate h–BN particles by various conditions, such as the rotation speed, the number of pass and the distance between the two rotation disks. We investigated the exfoliation effect of the WRD method by evaluating the statements of exfoliated h-BN particles from viewpoints of lateral size and thickness.

EXPERIMENTAL

A commercially available hexagonal boron nitride (h–BN, UHP–1, Showa Denko, Tokyo, Japan) was used as the laminated ceramic particles in this work. Aqueous h-BN slurry was prepared by h–BN particles and distilled water. The solid loading of h–BN particles in slurry was adjusted to be 0.5 or 2.0 vol.%. Aqueous h–BN slurries were prepared by the WRD mill system (Disperizer CDMX-T, Sintokogio, Aichi, Japan) as shown in Fig. 1. The rotation disk (100 mm diameter), which was made from alumina-sprayed coating stainless steel was used in order to prevent contamination during milling. The slurry was introduced between the parallel desks in the apparatus by a tube pump for controlling the flow rate. The rotating disk speed, the clearance which was the distance between two disks, and number of flow pass were adjusted at 3000~12000 rpm, 0.10~0.20 mm, and 1~10 pass, respectively.

The sedimentation tests of the WRD milling-assisted h–BN slurries were performed in borosilicate glass tube and allowed them to stand for 24 h. After 24 h, the sedimented cake heights of h–BN particles were measured. The initial slurry heights were adjusted to 50 mm. The sedimented cake height ratio was calculated by the ratio of the cake height to the slurry height. In order to study the characteristics of h-BN particles after the WRD process, particle size distributions were measured by a laser particle size analyzer (LA-920; HORIBA, Kyoto, Japan). The thickness of h-BN particles was evaluated by a field emission scanning electron microscopy

(FE-SEM, S-4300; Hitachi, Ibaraki, Japan). From FE-SEM images of h-BN particles, the thickness distributions of h-BN particles after the WRD method were measured in this work.

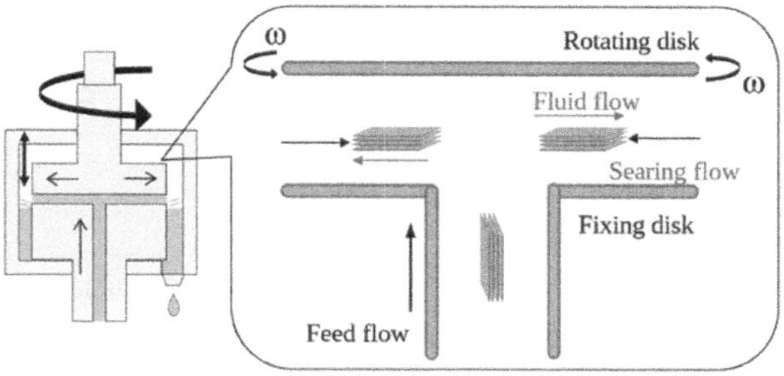

Figure 1. A diagram of wet-rotating disk milling system.

RESULTS AND DISCUSSION

Generally, it is known that the sedimentation test is a very important index for statement evaluation of particles such as pulverization, crush or exfoliation[24]. When the exfoliation of laminated h-BN particles is progressed, the sediments of h-BN particles become higher. Figure 2 shows the sedimented cake height ratios and median sizes (D_{50}) of h-BN particles by changing the number of flow pass as a function of the rotation speed. The solid loading, the flow rate, and the clearance were adjusted at 2.0 vol.%, 30 mL/sec, and 0.15 mm, respectively. In case of 1 flow pass, the sedimented cake height ratios of h-BN particles prepared at 3000, 6000, 12000 rpm of the rotation speed were 0.37, 0.42, 0.54, respectively. The increase of the sedimented cake height ratio was also observed by increasing the rotation speed when the flow pass was 5 passes. The sedimented cake height of the WRD milled h-BN particles became much higher with increasing the rotation speed. Furthermore, the repeat of the flow pass led to further the increase of the sedimented cake height ratio of h-BN particles. On the other hand, the sedimented cake height ratio of raw h-BN particles was higher as compared to those prepared at 3000 rpm. This result indicated that non-milled h-BN particles were agglomerated in slurry. Also, the median sizes of h-BN particles prepared by the WRD were lower as compared to raw h-BN particles in all cases. Especially, the median size of h-BN particles prepared by 10 passes was greatly decreased. These results indicated that the rotating disk method led to pulverization of h-BN particles not only exfoliation of them in this WRD condition.

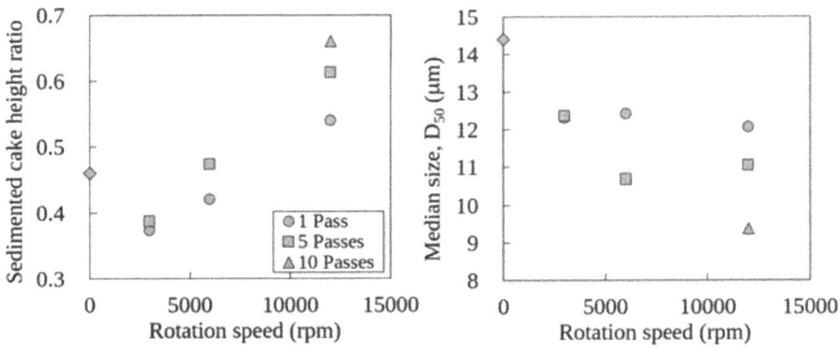

Figure 2. Sedimented cake height ratio (left) and median size (right) of the exfoliated h-BN particles as a function of the rotation speed.

In order to investigate the exfoliation behavior based on the parameter of the WRD, h-BN particles were prepared by changing the WRD conditions. Figure 3 shows the sedimented cake height ratios and median sizes of h-BN particles as a function of the flow rate. The solid loading, the rotation speed, the flow pass, and the clearance were adjusted to 2.0 vol.%, 12000 rpm, 5 passes, and 0.15 mm, respectively. As decreasing the flow rate, the sedimented cake height ratios and the median sizes were higher. This result indicated that the speed of h-BN particles fed to a rotating disk was very important for exfoliation of laminated particles. When h-BN particles were fed to the rotating disk too fast, h-BN particles would be not oriented between the rotating disks. So, the shearing flow would apply to h-BN particles from the random direction. Therefore, h-BN particles were destroyed by shearing flow. Table 1 also shows the median sizes of h-BN slurry containing 0.5 vol.% of h-BN particles by fixing the flow rate and the flow pass to 10 mL/sec and 10 passes, respectively. The median sizes of all h-BN samples prepared by the WRD were similar as compared to raw h-BN particles. Because the contacts between h-BN particles were reduced by decreasing the solid loading, the pulverization of h-BN particles was suppressed. Furthermore, the ratio of the sedimented cake height was also affected by the disk clearance during the milling. Figure 4 shows the sedimented cake height ratios h-BN particles by changing the disk clearance as a function of the rotation speed. The flow pass and the flow rate were performed at 5 passes and 10 mL/sec, respectively. The sedimented cake height ratios of h-BN particles were much higher with decreasing the disk clearance. Furthermore, the particle sizes of milled h-BN particles were almost similar with raw h-BN particles, meaning that h-BN particles were not crushed by this WRD condition. Thus, the sedimented cake behaviors of h-BN particles were different by the WRD conditions. In the WRD milling method, it is known that the strength of shearing flow is dependence on the parameters: rotation speed, disk clearance, and viscosity of fluid[25-27]. Therefore, the high shearing force generated by increasing the rotation speed and decreasing the disk clearance has effective to exfoliation of h-BN particles.

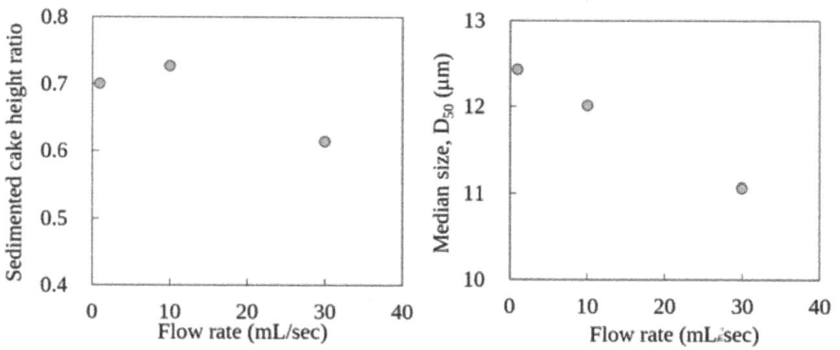

Figure 3. Sedimented cake height ratios (left) and median sizes (right) of h-BN particles as a function of the flow rate.

Table 1. Median sizes of h-BN slurry containing 0.5 vol.% of h-BN particles by fixing the flow rate and the flow pass to 10 mL/sec and 10 passes.

	Rotation speed (rpm)				
	0	3000	6000	9000	12000
Median size, D_{50} (μm)	14.4	13.5	13.3	13.7	14.0

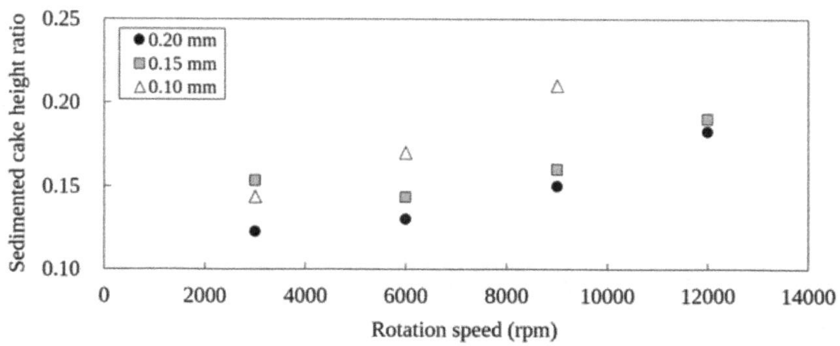

Figure 4. Sedimented cake height ratios of h-BN particles by changing the disk clearance as a function of rotation speed.

As shown in Fig.4, the sedimented cake height ratios of the WRD milled h-BN particles were different by the rotation speed and the disk clearance despite the particle sizes of them were almost similar. Therefore, the thickness of h-BN particles prepared by the WRD milling was investigated. Figure 5 represents FE-SEM images of h-BN samples. As shown in Fig. 5, the thicknesses of the WRD milled h-BN particles were thinner as compared to raw h-BN particles. These FE-SEM images show that the WRD milling led to exfoliation of h-BN particles. Furthermore, the 50% cumulative thickness (T_{50}) of h-BN particles exfoliated by 10 passes of flow pass at 0.20 mm of the disk clearance were obtained from about 150 thicknesses of FE-SEM images. Figure 6 shows the correlation between the sedimented cake height ratio and the 50% cumulative percent. The thickness revealed close correlation with the sedimented cake height ratio. The 50% cumulative percent of raw h-BN particles, the WRD-milled h-BN by rotating speed of 3000, 6000, 9000 and 12000 rpm were estimated at 0.170, 0.138, 0.130, 0.121 and 0.093 μm, respectively. The thicknesses of h-BN particles were much lower with increasing the rotating speed. These results showed that h-BN particles could be exfoliated by WRD method. Furthermore, as shown in Table 1, the particle sizes of h-BN particles were maintained. The average aspect ratios (D_{50}/T_{50}) were calculated from the mean particle size (D_{50}) and the mean thickness (T_{50}). The aspect ratios of h-BN particles of them were 85, 98, 102, 113, and 151, respectively. The average aspect ratio of milled h-BN particles was increased by up to 80% as compared to raw h-BN particles. Thus, WRD process leads to effective exfoliation of laminated h-BN particles without reducing the lateral sizes of h-BN particles. Finally, we will discuss the exfoliation mechanisms of h-BN by WRD method. In this process, h-BN particles in slurry are fed to the clearance between the rotating disks, and were oriented in the direction along the disk surface. As shown in Fig. 1, the high shearing force occurred by WRD is applied to the oriented h-BN particles from the edge direction. Therefore, the exfoliated h-BN particles with high aspect ratio are obtained by WRD method.

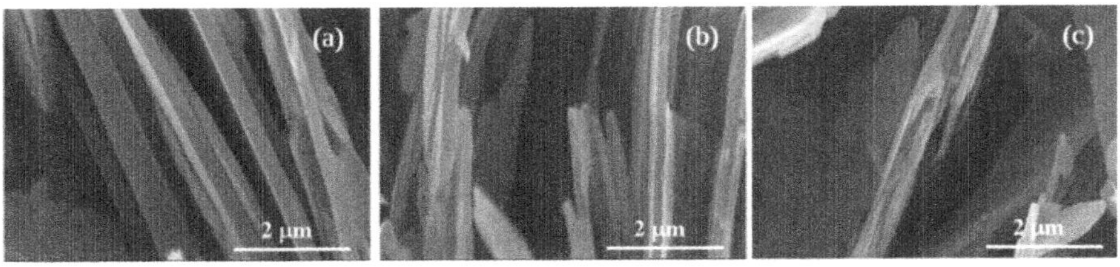

Figure 5. SEM images of slip-casted compacts observed from side direction.
(a) raw h-BN, (b) h-BN exfoliated by rotating speed of 9000 rpm or (c) 12000 rpm, 10 mL/sec of the flow rate, 15 mm of the disk clearance, 0.5 vol.% of the solid loading, and 5 passes of the flow.

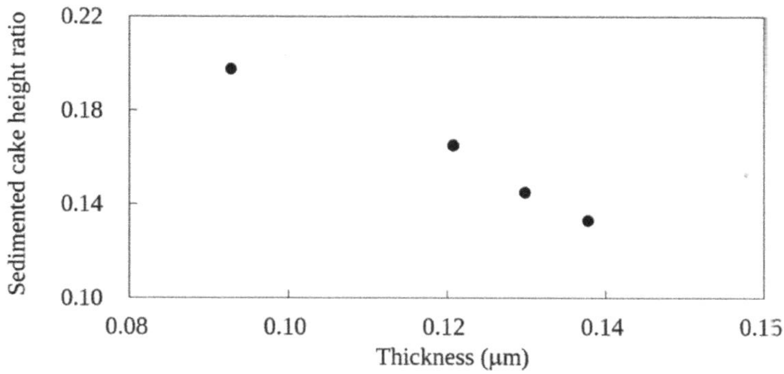

Figure 6. Correlation between the sedimented cake height ratio and the 50% cumulative percent, 10 mL/sec of the flow rate, 20 mm of the disk clearance, 0.5 vol.% of the solid loading, and 10 passes of the flow.

CONCLUSIONS

A wet-rotating disk method was the effective exfoliation method for laminated h-BN particle by controlling the conditions, such as the rotation speed, the flow pass, the disk clearance, the flow rate, and the solid loading. The lateral size and the thickness of h-BN particles were characterized by the sedimentation test, laser particle size analyzer and FE-SEM. The particle size of h-BN particle prepared by a WRD method was depended on the flow rate of h-BN slurry and the solid loading, resulting that the decreasing of lateral size of h-BN particles was suppressed with decreasing the flow rate and the solid loading. Moreover, the sedimented cake height ratios of h-BN particles obtained from the sedimentation test were increased with increasing the rotation speed, the flow pass process and narrowing the clearance between the disks. Also, the sedimented cake height ratios of h-BN showed the correlation with the thickness of h-BN particles measured by FE-SEM. These results suggested that the WRD milling led to exfoliation of h-BN particles. Thus, the WRD process leads to effective exfoliation of the laminated h-BN particles with high aspect ratio by controlling the WRD conditions.

REFERENCE

1. Geim, A. K. and Novoselov, K. S."The Rise of Graphene", Nat. Mater., 6 (3), pp. 183-191, 2007.

2. Giovannetti et al., "Substrate-Induced Band Gap in Graphene on Hexagonal Boron Nitride: Ab Initio Density Functional Calculations", Phys. Rev. B, 76 (7), pp. 073103, 2007.

3. Geim, A. K."Graphene: Status and Prospects", Science, 324 (5934), pp. 1530-1534, 2009.

4. Golberg et al., "Boron Nitride Nanotubes and Nanosheets", Acs Nano, 4 (6), pp. 2979-2993,

2010.
5. Song et al., "Large Scale Growth and Characterization of Atomic Hexagonal Boron Nitride Layers", Nano Lett., 10 (8), pp. 3209-3215, 2010.
6. Novoselov et al., "A Roadmap for Graphene", Nature, 490 (7419), pp. 192-200, 2012.
7. Wang et al., "Electronics and Optoelectronics of Two-Dimensional Transition Metal Dichalcogenides", Nat. Nanotechnol., 7 (11), pp. 699-712, 2012.
8. Van Khai et al., "Synthesis and Characterization of Single-and Few-Layer Mica Nanosheets by the Microwave-Assisted Solvothermal Approach", Nanotechnology, 24 (14), pp. 145602, 2013.
9. Cho et al., "Effects of H-Bn Additive on the Microstructure and Mechanical Properties of Aln-Based Machinable Ceramics", Materials Science and Engineering: A, 418 (1), pp. 61-67, 2006.
10. Li et al., "Machinable Al_2O_3/Bn Composite Ceramics with Strong Mechanical Properties", Mater. Res. Bull., 37 (8), pp. 1401-1409, 2002.
11. Rydberg et al., "Van Der Waals Density Functional for Layered Structures", Phys. Rev. Lett., 91 (12), pp. 126402, 2003.
12. Ren et al., "Large-Scale Production of Ultrathin Topological Insulator Bismuth Telluride Nanosheets by a Hydrothermal Intercalation and Exfoliation Route", J. Mater. Chem., 22 (11), pp. 4921-4926, 2012.
13. Coleman et al., "Two-Dimensional Nanosheets Produced by Liquid Exfoliation of Layered Materials", Science, 331 (6017), pp. 568-571, 2011.
14. Novoselov et al., "Electric Field Effect in Atomically Thin Carbon Films", Science, 306 (5696), pp. 666-669, 2004.
15. Nicolosi et al., "Liquid Exfoliation of Layered Materials", Science, 340 (6139), pp. 1226419, 2013.
16. Pacile et al., "The Two-Dimensional Phase of Boron Nitride: Few-Atomic-Layer Sheets and Suspended Membranes", Appl. Phys. Lett., 92 (13), pp. 133107, 2008.
17. Zhi et al., "Large-Scale Fabrication of Boron Nitride Nanosheets and Their Utilization in Polymeric Composites with Improved Thermal and Mechanical Properties", Adv. Mater., 21 (28), pp. 2889-2893, 2009.
18. Li et al., "Large-Scale Mechanical Peeling of Boron Nitride Nanosheets by Low-Energy Ball Milling", J. Mater. Chem., 21 (32), pp. 11862-11866, 2011.
19. Smith et al., "Large-Scale Exfoliation of Inorganic Layered Compounds in Aqueous Surfactant Solutions", Adv. Mater., 23 (34), pp. 3944-3948, 2011.
20. Lin et al., "Aqueous Dispersions of Few-Layered and Monolayered Hexagonal Boron Nitride Nanosheets from Sonication-Assisted Hydrolysis: Critical Role of Water", J. Phys. Chem. C, 115 (6), pp. 2679-2685, 2011.
21. Lu et al., "Carbon Nanofibres from Fructose Using a Light-Driven High-Temperature Spinning Disc Processor", Chem. Commun., 50 (12), pp. 1478-1480, 2014.
22. Enomura et al., "Development and Applicability of a New Medialess Disperser", J. Jpn. Soc.

Colour Mater., 75 (12), pp. 586-591, 2002.
23. Enomura et al., "Theoretical Analysis on the Flow inside a New Beadless Disperser", J. Jpn. Soc. Colour Mater., 77 (3), pp. 110-115, 2004.
24. Omura et al., "Fabrication of Stable Al_2O_3 Slurries and Dense Green Bodies Using Wet Jet Milling", J. Am. Ceram. Soc., 89 (9), pp. 2738-2743, 2006.
25. Iyer et al., "Continuous Flow Nano-Technology: Manipulating the Size, Shape, Agglomeration, Defects and Phases of Silver Nano-Particles", Lab Chip, 7 (12), pp. 1800-1805, 2007.
26. SwaminathanáIyer, K."Continuous Flow Nano-Technology: Manipulating the Size, Shape, Agglomeration, Defects and Phases of Silver Nano-Particles", Lab Chip, 7 (12), pp. 1800-1805, 2007.
27. Jachuck, R. and Ramshaw, C."Process Intensification: Heat Transfer Characteristics of Tailored Rotating Surfaces", Heat Recov. Syst. CHP, 14 (5), pp. 475-491, 1994.

Computational Design and Modeling

FEASIBLE AND RELIABLE *AB INITIO* APPROACH TO COMPUTATION OF MATERIALS RELEVANT FOR NUCLEAR WASTE MANAGEMENT

Piotr M. Kowalski[1], George Beridze[1], Yan Li[1], Yaqi Ji[1], Christoph Friedrich[2], Ersoy Şaşıoğlu[2] & Stefan Blügel[2]

[1]Institute of Energy and Climate Research: Nuclear Waste Management and Reactor Safety (IEK-6), Forschungszentrum Jülich GmbH, Jülich, Germany

[2]Peter Grünberg Institut and Institute for Advanced Simulation, Forschungszentrum Jülich GmbH and JARA, Jülich, Germany

Corresponding author: p.kowalski@fz-juelich.de

ABSTRACT

Nuclear waste generated during nuclear power production and other applications of nuclear technology requires safe storage and disposal. As a consequence, significant research effort is devoted to development of methods and materials that could be used for conditioning and disposal of long-lived radionuclides. We contribute to such research with atomistic simulations by providing a unique atomic-scale insight on the physics and chemistry of processes that determine the long-time behavior of radionuclide host materials. Such simulations are problematic, because density functional theory (DFT) - currently the only feasible *ab initio* method - fails dramatically for f-elements, including actinides. Therefore, using more resources demanding computational methods is proposed for calculations. This highly limits their applications to simplest molecular complexes and solids. However, we found that DFT+*U* method, with the Hubbard *U* parameter derived *ab initio* for each *f*-element, results in a good prediction of the structural and thermodynamic properties of *f*-materials such as uranium-bearing compounds, monazite-type ceramics (*(Ln,An)*PO_4) and pyrochlores ($Ln_2M_2O_7$), for which DFT often fails. We thus demonstrate that with carefully chosen *ab initio* methodology one can reliably simulate the properties of even chemically complex actinide-bearing materials, including these that are problematic or impossible to measure by the available experimental techniques. Last but not least we also discuss some results of application of *ab initio* molecular modeling to investigation of surface chemistry and diffusion of ^{14}C in (irradiated) graphite.

INTRODUCTION

Over the last two decades computational methods of quantum chemistry have been successfully applied to study minerals and fluids of chemically complex composition [1]. This was possible in part because of the tremendous increase in available computing power and subsequent development of complex quantum chemistry software. The most widely used method in computational materials science has been DFT [2]. This is because of its advantageous scaling comparing to other *ab initio* computations methods, which currently allows for *ab initio* computation and molecular dynamics simulations of hundreds of atoms. *Ab initio* modeling has been actively applied in nuclear science and nuclear waste management become a field especially attractive for computational science [3]. This is in part because it requires understanding of basic atomic scale structural and thermodynamic parameters of materials, including novel nuclear waste forms. However, because approximations used in DFT were designed to recover properties of light elements, the method fails, often even on qualitative level, for *f*-elements such as lanthanides and actinides. For instance it highly overestimates reaction enthalpies between uranium bearing molecular complexes and solids [4,5] and predicts incorrect electronic states for even simple materials like actinide-oxides [6]. Insulators such as UO_2, NpO_2 or PuO_2 are predicted to be metals [6]. As a consequence, more precise but computationally demanding, and often unfeasible *ab initio* methods such as hybrid functionals, MP2 or CCSD(T) are proposed for calculations of f-elements bearing systems [6,4]. This highly limits the applications of *ab initio* atomistic modeling to simplest molecular complexes and solids which prevents meaningful computer-aid atomistic modeling of chemically complex materials of interest for nuclear waste management, including solid solutions with actinides.

In our research we try to find a reliable and feasible computational approach which would permit *ab initio* simulations of chemically complex materials of interest for nuclear waste management. We applied DFT+*U* method, which is a simple modification of DFT allowing for direct accounting for correlations between *f* electrons [7]. This method has been widely applied in research on nuclear materials [8,9,10,11] but in those studies the Hubbard parameter *U*, which describes the on-site Coulomb repulsion between f-electrons, has been fixed to a usually constant value that in some cases was derived from the spectroscopic measurement (UO_2 [12,13]) but is usually guessed. In our research we apply the recently developed *ab initio* methods, the constrained local density approximation (cLDA) [13] and the constrained random phase approximation (cRPA) [14], in order to derive the value of the Hubbard *U* parameter and make the DFT+*U* method a parameter free approach. We will show that such an approach is crucial for correct computation of f-materials and allows for better understanding of the f-electrons driven chemistry of actinides. In addition to the application of DFT+*U* method we discuss the ways of treatment of f-electrons by pseudopotentials and the importance of spin-orbit effects. One commonly used technique is to not compute the f-electrons explicitly but to freeze them into the core and mimic their presence by the pseudopotential. We will show that such an approach is adequate for computing the excess properties of solid solutions.

With carefully chosen computational methodology we have performed simulations of various properties of nuclear materials. We will briefly discuss our results on derivation of structural and thermodynamic properties of monazite- and pyrochlore-type ceramics that are considered as novel nuclear waste forms and on the surface chemistry and diffusion processes in irradiated graphite.

COMPUTATIONAL APPROACH

The *ab inito* calculations of materials properties were performed with plane-wave DFT Quantum-ESPRESSO [15] code. We applied energy cutoff of 50Ryd, and the core electrons of considered elements were modeled by Vanderbilt pseudopotentials [16]. The calculations of the Hubbard *U* parameter were performed using the linear response approach of [7] (cLDA, see [17,5] for computational details) and the constrained random phase approximation [14] (cRPA). Using the cRPA methodology one defines an effective Hubbard *U* interaction between the localized electrons after restricting the screening to the ones which are not explicitly treated in the effective model Hamiltonian. The task of identifying transitions which are to be eliminated is sometimes complicated due to the mixing of *f* states with the s and p states. Derivation of the Hubbard *U* parameter in cRPA strongly depends on the choice of the energy window in which the localized orbital basis (Wannier orbitals [18] in this case) has to be formed [19]. Calculations in this work using the cRPA method were performed with the SPEX code [20] in which maximally localized Wannier functions (MLWFs)[21,22] are constructed internally and the FLEUR code [23] which uses the full-potential linearized augmented-plane-wave (FLAPW) method [24] for the initial ground state calculations.

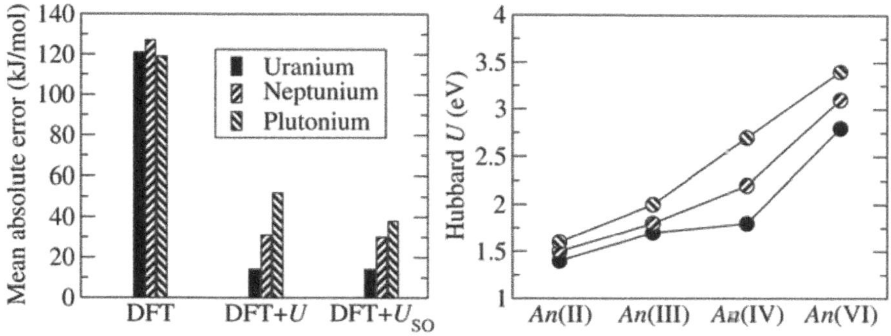

Figure 1. Left: The mean absolute error of the computed enthalpies of reactions involving actinide-fluorites (AnF_6, AnF_4, AnF_3 and AnF_2); Right: The average Hubbard U parameter value, derived for different actinide oxides, fluorites and chlorites, as a function of the oxidation state of the *An* cation.

The details on calculations of the monazite- and pyrochlore-type ceramics are provided in our previous studies [17,25,26,27]. The bulk graphite was modeled using a 2x2x2 supercell containing 128 atoms and the graphite surface was modeled with the system of the same size, i.e. 2x2 slab of four layers separated by 13.4 Å thick vacuum layer. In these calculations Van der Waals interactions were modeled using scheme of [28,29] as implemented in Quantum-ESPRESSO code.

RESULTS & DISCUSSION

REACTION ENTHALPES BETWEEN *An* BEARING COMPLEXES AND STRUCTURES OF SOLIDS WITH DFT+U METHOD

In our recent studies [5] we have shown that the DFT+U method with the Hubbard U parameter derived *ab initio* using the linear response method [7] results in significant improvement of the prediction of enthalpies of reactions between uranium-bearing molecules and solids. Here we extended those calculations to other actinides, namely Np and Pu. As illustrated in Figure 1 for these actinides the DFT+U results in similar to the case of uranium reduction of the mean absolute error of the enthalpies of reactions between various actinide-fuorides, bringing the uncertainties close to the experimental level of about 20 kJ/mol. It is known that for Np and Pu the spin-orbit interaction is more significant than in the case of uranium and our calculations show that it can affect the reaction enthalpies by as much as 50 kJ/mol (less than 2 kJ/mol for uranium). Therefore, we computed the spin-orbit correction for the cases of Np and Pu and the error of the computed reaction enthalpies with spin-orbit interaction accounted for is also given in Figure 1 (DFT+U_{SO}). On the average the spin-orbit corrections cancel out and results of both calculations, with and without spin-orbit, are similar, although accounting for the spin-orbit interaction reduces the error slightly. Similarly to the case of uranium [5] we found that the Hubbard U parameter depends on the oxidation state of an ancinide, which is clearly visible in Figure 1 (right panel). The Hubbard U parameter values decrease with decreasing the oxidation state of an *An* atom and the trend is clearly visible for all the considered actinides. This result puts into question the usage of a fixed value of the Hubbard U parameter for calculations of different actinide-bearing materials using DFT+U method.

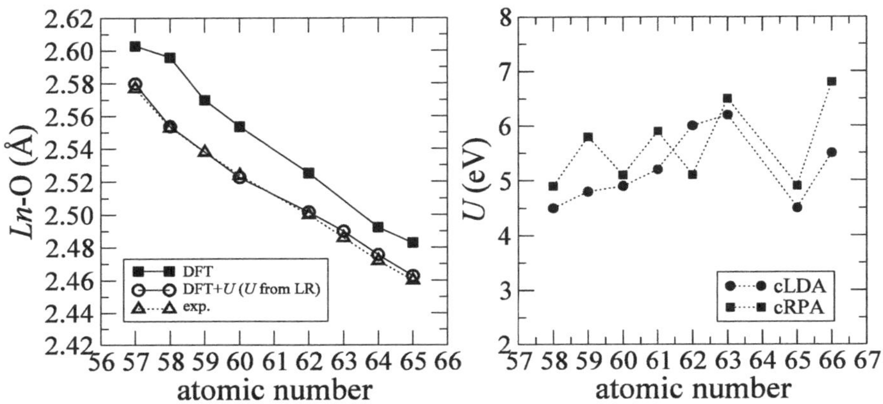

Figure 2. Left: *Ln*-O bond length predicted by DFT (PBE) [32], DFT+U with the Hubbard U parameter derived using the linear response method of [7] and measured experimentally. Right: the Hubbard U parameter computed with cLDA and cRPA methods for Ln_2O_3.

In subsequent calculations of ceramic waste forms we found that DFT+U method performs exceptionally well for structural parameters of monazite-type lanthanide-phosphates [17]. As illustrated in Figure 2 (left panel) usage of PBEsol exchange-correlation functional [30] and the Hubbard U parameter derived from linear response method [7] results in a nearly perfect reproduction of the *Ln*-O bond lengths in monazite and improves the prediction of commonly used PBE functional [31,32]. This successful result was achieved by computation of the Hubbard U parameters for each considered cations and structure, which is broadly discussed in [17]. There is a discussion in the literature whenever the two *ab initio* methods used for calculation of the Hubbard U parameter value, cLDA (e.g. linear response method of [7]) and cRPA give consistent results. Significant discrepancies have been observed for the elements at the end of 3*d* series [14]. Having obtained excellent results with the linear response method (cLDA) we recomputed the values of the Hubbard U parameters for Ln_2O_3 using cRPA. The comparison of the results of the two methods is given in Figure 2 (right panel). The two calculations give very consistent results with a deviation not larger than 1 eV. This shows that the both methods can be successfully utilized in computation of the Hubbard U parameter value for the lanthanide-bearing materials.

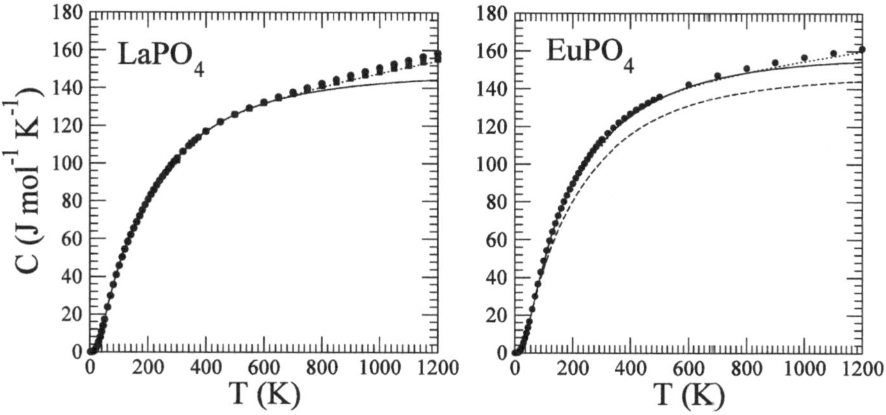

Figure 3. Heat capacities of $LaPO_4$ and $EuPO_4$ monazites. The lines represent the calculations using quasiharmonic approximation (solid), contribution from the lattice vibrations only (dashed) and results corrected for anharmonic effects (dotted). Points represent the experimental data. All the data are taken from [10] and references cited in that paper. The insert shows the low temperature difference between heat capacities of $EuPO_4$ and $LaPO_4$.

THERMODYNAMIC PARAMETERS OF MONAZITE-TYPE CERAMICS: EXCESS ENTHALPIES OF MIXING AND HEAT CAPACITIES

Thermodynamic parameters of materials considered for nuclear waste management are very important for the assessment of long term stability of novel waste forms. Two of the important parameters are the heat capacities and the excess enthalpy of mixing. In Figure 3 we present the modeled heat capacities of $LaPO_4$ and $EuPO_4$ monazites and the low temperature difference of the two. The computational results match well the observed values which permits analysis of the contributions to the heat capacity and its variation along the lanhanide series [25]. There are two contributions to the heat capacity of monazite, the lattice vibrations and the thermal excitation of *f* electrons (Schottky contribution) [33]. Because La^{3+} has no f electrons its heat capacity is given only by the lattice vibrations. On the other hand, in case of $EuPO_4$ there is significant Schottky contribution, which explains why it has higher heat capacity than $LaPO_4$. With the calculations we are able to explain the low temperature deep observed in the difference of the heat capacities of Eu and La monazites [34]. It results from the slightly smaller lattice vibrations contribution to the heat capacity of $EuPO_4$ (see insert in Fig. 3). By obtaining a good match to the experimental values for all the measured $LnPO_4$ cases we were able to explain the variation of the heat capacities between the different lanthanide-monazites and to make predictions for actinide-monazites [25].

Another parameter important in the assessment of the stability of a solid nuclear waste form is the excess enthalpy of mixing described by the interaction parameter [35]. It is known from the previous experimental studies [36] that the monazite-type solid solutions are well described by the regular model, so the excess enthalpy of mixing $H^E = Wx(1-x)$, where W is the Margules interaction parameter and *x* is the mole fraction of the substituting cations. In [26] we derived these interaction parameters for a series of (La,(Ln,An))PO_4 solid solutions and obtained a nice relationship between the strength of the interaction parameter and the mismatch in volume of endmembers. For (La,Eu)PO_4 solid solution we obtained the value of interaction parameter of 13.5 kJ/mol, which is consistent with the previous experimental estimate of 13.9±8.3 kJ/mol [36]. Knowing the value of the interaction parameter we can make prediction regarding stability of this solid solution. The Margules interaction parameter value of 13.5 kJ/mol results in the solid solution being stable in the entire range for temperature above 800 K. Below this temperature a wide miscibility gap forms, which is indicated in the Figure 4. This result suggests that such a solid solution is unstable at ambient conditions and the resulting monazite would likely represents a metastable phase if synthesized. In a similar way one can make assessment of the stability of the monazite-type solid solutions with actinides [26]. We note that in these calculations we used the *f*-in-core pseudopotentials approach. This is justified because considering small differences between energies of isostructuraly identical materials, all the effects related to strong correlations of *f*-elecrons cancel out to significant extend.

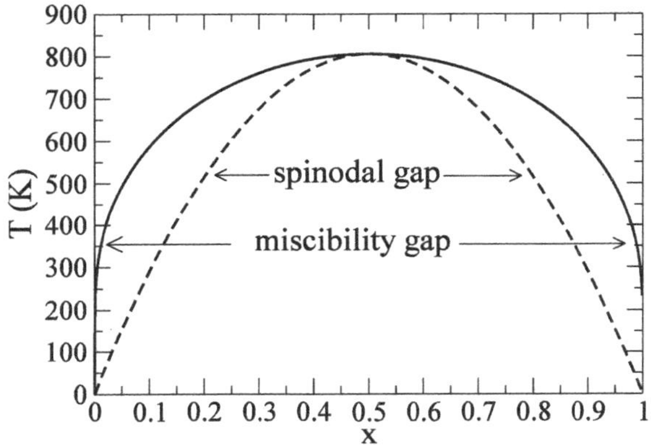

Figure 4. The miscibility gap predicted for $La_{1-x}Eu_xPO_4$ solid solution.

DEFECT FORMATION ENERGIES IN $Ln_2Zr_2O_7$ PYROCHLORES

Pyrochlore-type ceramics are potential candidates for a final nuclear waste disposal form because certain pyrochlores under irradiation transfer to the disordered fluorite instead of becoming amorpous [37]. In order to understand this phenomenon it is important to understand the origin of the order-disorder transition in pyrochlores. The first step is to determine the energetics of defects formation in pyrochlore. We thus performed the *ab initio* calculations of cation-antisite and anion Frenkel pair defect formation energies in pyrochlores [27]. The results for $Ln_2Zr_2O_7$ are provided in Figure 5 together with the results of previous atomistic modeling that used simple force-fields approach [38,39]. It is clear that the previous calculations highly overestimated the defect formation energies by up to 6eV. Also on the qualitative level, *ab initio* calculations reveal different trends in the defect formation energies along the lanthanide series.

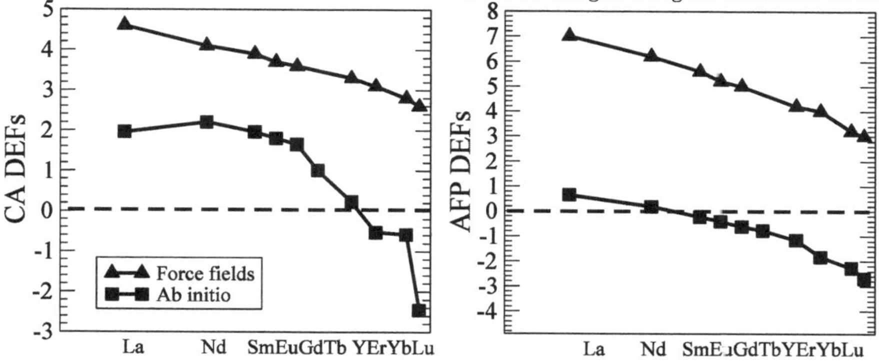

Figure 5. The force field [38,39] and *ab initio* prediction of the cation antisite (CA) and anion Frenkel pair (AFP) defect formation energies (DFEs) in $Ln_2Zr_2O_7$ pyrochlores.

Interestingly, both approaches, DFT+*U* and f-in-core, resulted in very similar energetics (within 0.3 eV), which permits using more stable f-in-core approach for all the calculations. The obtained defect formation energies correlate well with the stability field of pyrochlores. We found that the compositions that form disordered fluorite have negative anion Frenkel pair defect formation energies. This indicates that diffusion of the oxygen atoms is one of the factors leading to the disordering of selected pyrochlores and their subsequent transition to the disordered fluorite phase. More information on these studies are provided in [27] and in [40] (this volume)

SURFACE CHEMISTRY AND DIFFUSION IN I-GRAPHITE

Besides actinides, ^{14}C containing irradiated graphite represents significant amount of radioactive waste generated during operations of nuclear reactors. Safe disposal strategies of nuclear graphite require understanding of surface reactivities, radiocarbon ^{14}C diffusion and its potential release mechanisms. These topics could be successfully investigated by atomistic modeling. We performed atomistic modeling studies of the interaction of carbon surface with water molecules and the diffusion of ^{14}C in graphite. In line with previous works we found that water molecule is weakly bounded to the surface of ideal graphite (bonding energy of 0.037 eV), but easily undergoes dissociation on the defect sites [41]. Representative adsorption geometries for dissociation of water on the surface vacancy sites are given in Figure 6. The adsorption on mono-vacancy and double-vacancy result in 3.4 eV and 4.7 eV adsorption energies. These are consistent with previous studies [41]. This could potentially explain hydrophilic character of irradiated graphite as opposite to the hydrophobic behavior of normal graphite.

One of the interesting topics is the self diffusion of ^{14}C in graphite. There are three main diffusion mechanisms considered: (1) diffusion through vacancy migration, (2) diffusion through direct interchange and (3) diffusion of the interstitial atoms. Energetics of these processes has been considered by a simple modeling using estimates of C-C bond strengths [42]. The obtained energies were 8.3 eV, 3.9 eV and 18.1 eV for the processes 1, 2 and 3 respectively. We have recomputed these activation energies using DFT and got 16.1 eV, 9.5 eV and 15.6 eV, respectively.

These energies are substantially larger than the measured activation barrier for diffusion of 6.9 eV [43], although this suggests that the direct interchange is the most energetically favorable process that leads to the selfdiffusion of ^{14}C. Interestingly, the main contributors to the 1 and 2 processes are the large vacancy and interstitial formation energies, 7.6 eV and 7.4 eV, respectively. The barriers for diffusion of a surface, interstitial atom and the vacancy are much smaller, 0.6 eV and 1.1 eV respectively. These preliminary results show a need for more studies of i-graphite in order to understand properties of this material and its behavior under disposal conditions.

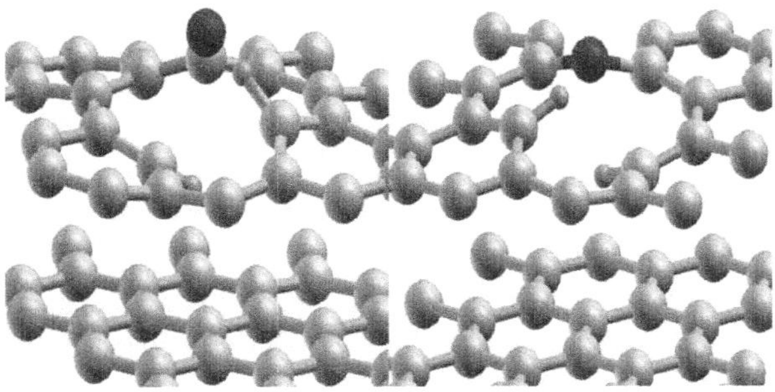

Figure 6. The configurations of water dissociated on mono-vacancy (left) and double-vacancy (right) defects on the graphite surface.

CONCLUSIONS

Here we have presented the selected results of the *ab initio* investigation of the actinide- and lanthanide-bearing materials and graphite. We show that with an appropriate choice of the computational method the structural and thermodynamic properties of these materials can be computed with computationally inexpensive extensions of DFT such as DFT+*U*. We discussed significant reduction of the error of the predicted reaction enthalpies computed with DFT+*U* method, when the Hubbard *U* parameter is computed *ab initio.* We found that the value of the Hubbard *U* parameter depends on the oxidation state of actinide-cation. We showed that the structural parameters of different monazite-type ceramics can be also very accurately predicted with the DFT+*U* method. Interestingly, we checked that the two methods used for calculations of the Hubbard *U* parameters, cLDA and cRPA, give consistent values for Ln_2O_3. In subsequent examples of applications of the *ab initio* simulations to nuclear materials we discussed the results of our studies of the heat capacities and the excess enthalpies of mixing of monazite-type solids and solid solutions and the energetics of defect formation in pyrochlores. Finally, we showed two examples of how *ab initio* simulations can improve our understanding of the surface chemistry and ^{14}C diffusion in irradiated graphite. This was done through consideration of water adsorption on ideal and defective graphite surfaces and comparing the energetics of different diffusion processes of radiocarbon. All the presented results show the usefulness of modern *ab initio* based computational techniques in investigation of both computationally and experimentally challenging nuclear materials.

ACKNOWLEDGEMENTS
We acknowledge the computing time on RWTH Aachen cluster awarded through Jülich-Aachen research alliance (JARA-HPC).

REFERENCES

[1] Jahn, S., & Kowalski, P. M. (2014). Theoretical Approaches to Structure and Spectroscopy of Earth Materials. *Reviews in Mineralogy and Geochemistry, Volume 78,* 691-743.

[2] Koch, W., & Holthausen, M. C. (2000). A Chemist's Guide to Density Functional Theory. Wiley-VCH, Weinheim.

[3] Chrones, A., Rushton, M. J. D., Jiang, C., & Tsoukalas, L. H. (2013). Nuclear Wasteform Materials: Atomistic Simulation Case Studies. *Journal of Nuclear Materials, Volume 441,* 29-39.

[4] Shamov, G. A., Schreckenbach, G., & Vo, T. N. (2007). A Comparative Relativistic DFT and Ab Initio Study on the Structure and Thermodynamics of the Oxofluorides of Uranium(IV), (V) and (VI). *Chemistry – A European Journal, Volume 13,* 4932-4947.

[5] Beridze, G., & Kowalski, P. M. (2014). Benchmarking the DFT+U Method for Thermochemical Calculations of Uranium Molecular Compounds and Solids. *Journal of Physical Chemistry A, Volume 118,* 11797-11810.

[6] Wen, X.D., Martin, R. L., Henderson, T. M., & Scuseria, G. E. (2013). Density Functional Theory Studies of the Electronic Structure of Solid State Actinide Oxides. *Chemical Reviews, Volume 113,* 1063-1096.

[7] Cococcioni, M., & de Gironcoli, S. (2005). Linear Response Approach to the Calculation of the Effective Interaction Parameters in the LDA+U Method. *Physical Review B, Volume 71,* 035105.

[8] Dorado, B., Amadon, B., Freyss, M., & Bertolus, M. (2009). DFT+U Calculations of the Ground State and Metastable States of Uranium Dioxide. *Physical Review B, Volume 79,* 235125.

[9] Dudarev, S. L., Manh, D. N., & Sutton, A. P. (1997). Effect of Mott- Hubbard Correlations on the Electronic Structure and Structural Stability of Uranium Dioxide. *Philosophical Magazine Part B, Volume 75,* 613–628.

[10] Rak, Z., Ewing, R. C., & Becker, U. (2011). Role of Iron in the Incorporation of Uranium in Ferric Garnet Matrices. *Physical Review B, Volume 84,* 155128.

[11] Odoh, S. O., Pan, Q. -J., Shamov, G. A., Wang, F., Fayek, M., & Schreckenbach, G. (2012). Theoretical Study of the Reduction of Uranium- (VI) Aquo Complexes on Titania Particles and by Alcohols. *Chemistry - A European Journal, Volume 18,* 7117–7127.

[12] Kotani, A., & Yamazaki, T. (1992). Systematic Analysis of Core Photoemission Spectra for Actinide Di-Oxides and Rare-Earth Sesqui-Oxides. *Progress of Theoretical Physics Supplement, Volume 108,* 117–131.

[13] Baer, Y., & Schoenes, J. (1980). Electronic Structure and Coulomb Correlation Energy in UO_2 Single Crystal. *Solid State Communications, Volume 33,* 885–888.

[14] Aryasetiawan, F., Karlsson, K., Jepsen, O., & Schönberger, U. (2006). Calculations of Hubbard U from first-principles. *Physical Reveiw B, Volume 74,* 125106.

[15] Giannozzi, P., Baroni, S., Bonini, N., Calandra, M., Car, R., Cavazzoni, C., Ceresoli, D., Chiarotti, G. L., Cococcioni, M., Dabo, I., et al. (2009). QUANTUM ESPRESSO: a Modular and Open-Source Software Project for Quantum Simulations of Materials. *Journal of Physics: Condensed Matter, Volume 21,* 395502.

[16] Vanderbilt, D. (1990). Soft Self-consistent Pseudopotentials in a Generalized Eigenvalue Formalism. *Physical Review B, Volume 41,* 7892.

[17] Blanca-Romero, A., Kowalski, P. M., Beridze, G., Schlenz, H., & Bosbach, D. (2014). Performance of DFT+U Method for Prediction of Structural and Thermodynamic Parameters of Monazite-type Ceramics. *Journal of Computational Chemistry, Volume 35,* 1339-1346.

[18] Wannier, G. H. (1937). The Structure of Electronic Excitation Levels in Insulating Crystals. *Physical Review, Volume 52,* 191-197.

[19] Şaşıoğlu, E., Friedrich, C., & Blügel S. (2011). Effective Coulomb Interaction in Transition Metals from Constrained Random-phase Approximation. *Physical Review B, Volume 83,* 121101.

[20] Friedrich, C., Blügel, S., & Schindlmayr, A. (2010). Efficient Implementation of the GW Approximation Within the All-electron FLAPW Method. *Physical Review B, Volume 81,* 125102.

[21] Marzari, N., & Vanderbilt, D. (1997). Maximally Localized Generalized Wannier Functions for Composite Energy Bands. *Physical Review B, Volume 56,* 12847 – 12865.

[22] Mostofi, A. A., Yates, J. R., Lee, Y. -S., Souza, I., Vanderbilt, D., & Marzari, N. (2008). Wannier90: A Tool for Obtaining Maximally-Localised Wannier FunctionsComput. Computer Physics Communications, Volume 178, 685-699.

[23] www.flapw.de

[24] Wimmer, E., Krakauer, H., Weinert, M., & Freeman, A. J. (1981). Full-potential Self-consistent Linearized-augmented-plane-wave Method for Calculating the Electronic Structure of Molecules and Surfaces: O_2 molecule. *Physical Review B, Volume 24,* 864-875.

[25] Kowalski, P. M., Beridze, G., Vinograd, V. L., & Bosbach, D. (2015). Heat Capacity of Lanthanide and Actinide Monazite-type Ceramics. *Journal of Nuclear Energy, Volume 464,* 147-154.

[26] Li, Y., Kowalski, P. M., Blanca-Romero, A., Vinograd, V., & Bosbach, D. (2014). Ab Initio Calculation of Excess Properties of Solid Solutions. *Journal of Solid State Chemistry, Volume 220,* 137-141.

[27] Li, Y., Kowalski, P. M., Beridze G., Birnie, A., Finkeldei, S., & Bosbach, D. (2015). Defect Formation Energies in $A_2B_2O_7$ Pyrochlores. *Scripta Materialia,* doi:10.1016/j.scriptamat.2015.05.010.

[28] Barone, V., Casarin, M., Forrer, D., Pavone, M., Sambi M., & Vittadini, A. (2009). Role and Effective Treatment of Dispersive Forces in Materials: Polyethylene and Graphite Crystals as Test Cases. *Journal of Computational Chemistry, Volume 30,* 934-939.

[29] Grimme, S. (2006). Semiempirical GGA-type Density Functional Constructed with a Long-range Dispersion Correction. *Journal of Computational Chemistry, Volume* 27, 1787-1799.

[30] Perdew, J. P., Ruzsinszky, A., Csonka, G. I., Vydrov, O. A., Scuseria, G. E., Constantin, L. A., Zhou, X., & Burke, K. (2008). Restoring the Density-Gradient Expansion for Exchange in Solids and Surfaces. *Physical Review Letters, Volume 100,* 136406.

[31] Perdew, J. P., Burke, K., & Ernzerhof, M. (1996). Generalized Gradient Approximation Made Simple. *Physical Review Letters, Volume 77,* 3865.

[32] Rustad, J. R. (2012). Density Functional Calculations of the Enthalpies of Formation of Rare-earth Orthophosphates. *American Mineralogist, Volume 97,* 791–799.

[33] Westrum, E. F. Jr. (1985). Schottky Contributions in Chemical Thermodynamics. *Journal of Thermal Analysis and Calorimetry, Volume 30,* 1209–1215.

[34] Thust, A., Arinicheva, Haussuehl, Y. E., Bayarjagal, L., Vogel, S. C., Neumeier, S., & Winkler, B. (2015). *Journal of the American Ceramic Society,* (submitted for publication).

[35] Glynn, P. (2000). Solid-Solution Solubilities and Thermodynamics: Sulfates, Carbonates and Halides. *Reviews in Mineralogy and Geochemistry, Volume 40,* 481–511.

[36] Popa, K., Konings, R. J. M., & Geisler, T. (2007). High-temperature Calorimetry of $(La_{1-x}Ln_x)PO_4$ Solid Solutions. *Journal of Chemical Thermodynamics, Volume 39,* 236–239.

[37] Ewing, R. C., Weber, W. J., & Lian, J. (2004). Nuclear Waste Disposal—Pyrochlore Nuclear Waste Form for the Immobilization of Plutonium and "Minor" Actinides. *Journal of Applied Physics, Volume 95,* 5949.

[38] Sickafus, K. E., Minervini, L., Grimes, R. W., Valdez, J. A., Ishimaru, M., Li, F., McClellan, K. J., & Hartmann, T. (2000). Radiation Tolerance of Complex Oxides. *Science, Volume 289,* 748–751.

[39] Minervini, L., Grimes, R. W., & Sickafus, K. E. (2000). Disorder in Pyrochlore Oxides. *Journal of the American Ceramic Society, Volume 83,* 1873–1878.

[40] Li, Y., Kowalski, P. M., Beridze, G., & Ji, Y. (2015). *Ceramic Transactions* (this volume)

[41] Cabrera-Sanfelix, P., & Darling, G. R. (2007). Dissociative Adsorption of Water at Vacancy Defects in Graphite. *Journal of Physical Chemistry C, Volume 111,* 18258-18263.

[42] Dienes, G. J. (1952). Mechanism for Self-Diffusion in Graphite. *Journal of Applied Physics, Volume 23,* 1194-1200.

[43] Kanter, M. A. (1957). Diffusion of Carbon Atoms in Natural Graphite Crystals. *Physical Review, Volume 107,* 655.

Materials for Extreme Environments

PHASE EVOLUTION PHENOMENON DURING HOT PRESSING OF THE SHS OBTAINED Ti_3AlC_2 PRECURSORS POWDERS

L. Chlubny[1], J. Lis[1], Cz. Kapusta[2], D. Zientara[1], K. Chabior[1], P. Chachlowska[1]
[1]AGH - University of Science and Technology, Faculty of Material Science and Ceramics Department of Technology of Ceramics and Refractories, Al. Mickiewicza 30, 30-059 Cracow, Poland
[2]AGH University of Science and Technology, Faculty of Physics and Applied Computer Science, Al. Mickiewicza 30, 30-059 Kraków, Poland

ABSTRACT

Ti_3AlC_2 is one of the many compounds which belong to the very interesting group of materials called MAX phases. The MAX phases are heterodesmic layered materials with different chemical bonds which strongly influence their properties locating them between metals and ceramics. They are characterised with high elastic modulus, low hardness, very high fracture toughness and high electrical and heat conductivity which leads to many potential application as a construction or functional materials.

One of the most effective and efficient methods of synthesizing MAX phase precursors powders is Self-propagating High-temperature Synthesis (SHS). This method is based on utilization of exothermal effect of reaction and allows obtaining numerous materials in relatively short times and at low energy consumption.

The SHS was applied for manufacturing of Ti_3AlC_2 active precursors powders containing high volume of ternary layered materials. During the densification of such obtained powders by hot-pressing process (HP) further chemical reactions were observed in sintered powders. As the result of these processes the phase evolution phenomenon of sintered materials was observed and near single phase materials were obtained. The changes of apparent densities in various sintering conditions were examined.

INTRODUCTION

In the Ti-Al-C-N system beside many covalent materials, such as nitrides and carbides, a group of ternary and quaternary layered compounds can be found. These materials are referred in literature as MAX phases, H-phases, Novotny-phases or thermodynamically stable nanolaminates. The most popular term for these compounds MAX phases is related to their specific $M_{n+1}AX_n$ stoichiometry, where M is an early transition metal, A is an element of A groups (mostly IIIA or IVA) and X is carbon and/or nitrogen. Heterodesmic structures of these phases are hexagonal, P63/mmc, and specifically layered. They consist of alternate near close-packed layers of M_6X octahedrons with strong covalent bonds and layers of A atoms located at the centre of trigonal prisms. The M_6X octahedral, similar to those forming respective binary carbides, are connected one to another by shared edges. Variability of chemical composition of the nanolaminate is usually labeled by the symbol describing their stoichiometry, e.g. Ti_2AlC represents 211 type phase and Ti_3AlC_2 – 312 type. Differences between the respective phases consist in the number of M layers separating the A-layers: in the 211's there are two whereas in the 321's three M-layers [1-3]. This layered, heterodesmic structure of MAX phases leads to the unique set of properties. These materials combine some features characteristic for ceramics such as high stiffness, moderately low thermal expansion coefficient and excellent thermal and chemical resistance with low hardness, good compressive strength, high fracture toughness, ductile behavior, good electrical and thermal conductivity which are typical for metals. They can be applied as a part of ceramic armour based on functionally graded materials (FGM), as a matrix in ceramic-based composites reinforced by covalent phases or as functional materials due to their thermal and electrical properties [4, 5]. The wide spectrum of information about properties, synthesis and applications of MAX phases materials in Ti-Al-C system can be found in complex and comprehensive review by X.H. Wang and Y.C. Zhou [6] and latest works of M.W. Barsoum [7] and M. Radovic [8].

The Self-propagating High-temperature Synthesis (SHS) is a method applied for obtaining over 1000 materials such as carbides, borides, nitrides, oxides, intermetallic compounds and composites. It appears also in the synthesis of advanced materials such as biomaterials, nanomaterials or materials for space applications. The main principle of this method is utilization of exothermal effect of chemical synthesis. The second of the conditions necessary to carry out the synthesis is to provide adiabatic conditions in the reacting system. This synthesis can proceed in a powdered bed of solid substrates or in the filtration combustion regime where at least one of the substrates is in gaseous state. An external source of heat has to be used to initiate the process and then the self-sustaining reaction front is propagating through the bed of substrates until all of them are consumed. This process could be initiated by the local ignition or by the thermal explosion system. The final form of the synthesized material may depends on kind of precursors used for synthesis and the technique applied. The SHS method belongs to a group called "flash chemical synthesis" characterized by short reaction times and almost immediate obtaining of the expected compounds or precursors of materials. It is characterized by rapid and spontaneous course and one of the characteristic features and at the same time one of the biggest advantages, is the ability to freeze the nonequilibrium phase thereby preventing the diffusion processes, which may change the properties of the final product. The other advantages of this method are low energy consumption, high temperatures obtained during the process, high efficiency and simple apparatus. The lack of control of the process is the disadvantage of this method [9].

The possibility of obtaining powders with high content of Ti_3AlC_2 by SHS synthesis was proved by authors in their previous work, where use of excess intermetallic material, namely TiAl together with elemental titanium powder and carbon resulted with powder containing 76.4 wt% of Ti_3AlC_2, 18.9 wt. % of TiC and 4.7 wt. % of Al_4C_3 [10]. Also other authors are confirming possibilities of synthesis of powders or bulk material characterised with high purity of the MAX phase. In the paper by X.H. Wang and Y.C. Zhou a vast part concerning the state of art of latest Ti_2AlC and Ti_3AlC_2 synthesis method can be found [6]. The possibility of Ti_3AlC_2 synthesis was also reported by A.G. Zhou et.al in case of synthesis where elemental powders were used and mixture of both MAX phases was obtained as a result [11]. Possibility of using intermetallic materials in the MAX phases in Ti-Al-C system SHS synthesis were proved in paper of Lopacinski et.al [12]. Different system of ignition, namely thermal explosion was applied by Y. Khoptiar and I. Gotman [13]. Also works by Liu et.al provides new information about SHS synthesis of Ti_3AlC_2 [14].

The main goal of this work was to examine phase evolution phenomenon during hot-pressing of Ti_3AlC_2 active precursors powders obtained by SHS method in order to establish optimal hot pressing temperatures. It was intended to eliminate TiC impurities which strongly affect properties of the dense, sintered material, decreasing its pseudo-plastic behaviour. The influence of scaling up process on SHS of the powders was observed. Also densification of sintered material was observed by examination of apparent densities of obtained samples.

PREPARATION

Following the experience gained during previous synthesis of ternary materials such as Ti_3AlC_2, Ti_2AlN and also Ti_2AlC, intermetallic materials in the Ti-Al system, metallic titanium powder and graphite powder as a source of carbon were selected to be used as precursors for SHS synthesis of Ti_3AlC_2 powders [10, 15, 16, 17, 18, 19]. Due to the relatively low availability of fine commercial intermetallic powders in the Ti-Al system, it was decided to synthesize them in the first stage of researches. Basing on previous work on optimization of SHS synthesis of Ti_3AlC_2 TiAl and Ti_3Al powders were selected to be synthesized by SHS method [10, 17, 22]. Titanium powder (AEE TI-109, 99.7% pure, ~100 mesh) and metallic aluminium powder (Benda-Lutz Co., grain size below 6.7 μm, +99.7% pure) were used in the SHS synthesis. The mixtures for SHS reaction were set in stoichiometric proportions according to equation:

$$Ti + Al \rightarrow TiAl \quad (1)$$

$$3Ti + Al \rightarrow Ti_3Al \quad (2)$$

Mixtures of powders were initially homogenized in a ball-mill for 24 hours with WC balls. After that the homogenized mixture was placed in a graphite crucible which was heated in a graphite furnace up to 1200°C. At this temperature the thermal explosion in the whole volume of powder occurred and the SHS reaction was initiated. Then obtained products, which had form of dense material, were initially mechanically disintegrated and crushed in a roll crusher to the grain size ca. 1 mm. Such products were afterwards ground to the grain size ca. 10μm for 6 hours in the rotary-vibratory mill in dry isopropanol, using WC balls as a grinding medium. Then obtained powders were dried and used as substrates for synthesis of ternary MAX phase materials. Other powders used during synthesis of Ti_3AlC_2 were commercially available aluminium powder (Benda-Lutz Co., grain size below 6.7 μm, +99.7% pure), graphite powder used as a source of carbon (Merck no. 1.04206.9050, 99.8% pure, grain size 99.5% < 50μm), titanium powder (AEE TI-109, 99.7% pure, ~325 mesh).

All of the Ti_3AlC_2 syntheses were conducted by SHS method with a local ignition system and with use of various precursors' stoichiometry. Basing on a previous researches were smaller quantities of powders were used three reactions were selected for upscaling process [10, 22]. The selected SHS reactions are given below. The 1.1, 1.2 corresponds to 10% and 20% of excess precursor respectively due to the extremely high temperatures in SHS reaction front resulting in evaporation of part of the substarate.

$$TiAl + 2Ti + 2C \rightarrow Ti_3AlC_2 \quad (3)$$

$$1.1TiAl + 2Ti + 2C \rightarrow Ti_3AlC_2 \quad (4)$$

$$1.2Ti_3Al + Ti + C \rightarrow Ti_3AlC_2 \quad (5)$$

The mixtures were set in assumed stoichiometric proportions and homogenized for 24 hours by the same method which was applied for intermetallic materials. Up to 125g of homogenized substrates mixtures were placed in the SHS chamber and ignited by the local ignition system. The time of ignition varied from 30 to 60 second. After the synthesis obtained products were ground using the same method as it was applied for intermetallic materials. When the fine powders were achieved the XRD analysis method was applied to determine phase composition of the synthesised materials. The data for quantitative and qualitative phase analysis were acquired from ICCD [20]. Quantities of the respective phases were calculated according to the Rietveld analysis [21]. The measurements were made within an accuracy of 0.5%.

In the next step SHS derived powders were prepared for hot-pressing process. The active sintering precursor powders of Ti_3AlC_2 materials were placed in a 1 inch diameter graphite die and subjected to the hot-pressing process. The hot-pressing conditions were as following: maximum sintering temperatures were 1100, 1200 and 1300°C, time of annealing 1 hour, maximum pressure 25MPa. After hot-pressing and removing the graphite foil from the surfaces of the samples, obtained dense pellets were investigated by XRD method to study evolution of phase composition and also relative density was determined by the hydrostatic weighing.

RESULTS AND DISCUSSION

The X-ray diffraction analysis of SHS derived intermetallic powders TiAl and Ti_3Al proved that the first powder contains ca. 80% of TiAl and ca. 20% of Ti_3Al impurities while the second powder was 100% pure.

The XRD analysis of SHS derived active precursors of ternary MAX phase material Ti_3AlC_2 showed that upscaling process of his materials especially in the case of reaction 3 showed improvement and up-scalled powders were characterised by higher amount of ternary phase than it was in case of smaller samples[10]. Results of phase quantities analysis compared to the results of smaller samples are presented in Table 1.

Table I. Products of SHS synthesis of Ti_3AlC_2 phase composition.

Reaction No.	Weight of the sample [g]	Chemical reaction	Ti_3AlC_2 [wt.%]	Ti_2AlC [wt.%]	TiC [wt.%]	Al_4C_3 [wt.%]
3 [10]	25	$TiAl + 2Ti + 2C \rightarrow Ti_3AlC_2$	44.1	18.1	27.2	10.7
3	125	$TiAl + 2Ti + 2C \rightarrow Ti_3AlC_2$	70.9	7.7	21.4	-
4 [10]	25	$1.1\ TiAl + 2Ti + 2C \rightarrow Ti_3AlC_2$	76.4	18.9	4.7	-
4	125	$1.1\ TiAl + 2Ti + 2C \rightarrow Ti_3AlC_2$	74.3	9.1	16.7	-
5 [10]	25	$1.2Ti_3Al + Ti + C \rightarrow Ti_3AlC_2$	19,3	37.8	31.6	11.4
5	125	$1.2Ti_3Al + Ti + C \rightarrow Ti_3AlC_2$	10	64.3	25.7	-

In the case of the reaction 3 and 4, the amount of Ti_3AlC_2 phase is 70.9 and 74.3 wt.% respectively. This dominating phase was accompanied in both cases by the other MAX phase, namely Ti_2AlC and titanium carbide. In the case of reaction 5 significant amount of 211 phase Ti_2AlC was synthesized in the powder at favour of 312 phase Ti_3AlC_2, TiC and Al_4C_3. The XRD pattern of obtained powder with highest content of Ti_3AlC_2 is presented on Figure 1.

The hot-pressing process showed that SHS derived active sintering precursors are very good powders to obtain dense polycrystalline MAX phases material. Thanks to their properties and phase composition they are subjected to further chemical reaction during sintering resulting in obtaining almost single phase ternary materials. Higher hot pressing temperatures promote formation of Ti_3AlC_2 phase at cost of Ti_2AlC and TiC. This results in reduction of regular titanium carbide phase which may positively affect mechanical properties of these materials. The results of the phase evolution process during the hot-pressing are presented in Table II-IV.

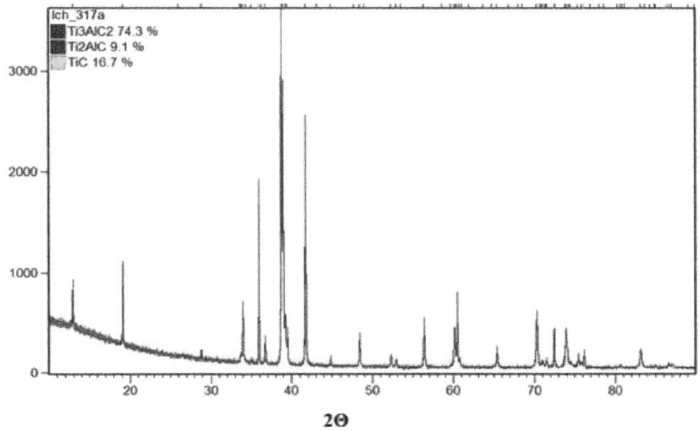

Figure 1. XRD pattern of the Ti_3AlC_2 powders obtained by SHS reaction 4

Table II. Phase evolution of hot-pressed SHS derived Ti_3AlC_2 powder obtained as a result of synthesis TiAl + 2Ti + 2C.

Sintering temperature[°C]	Ti_2AlC	Ti_3AlC_2	TiC
20	7,7	70,9	21,4
1100	4,5	77,2	18,3
1200	-	84,4	15,6
1300	-	84,8	15,2

Table III. Phase evolution of hot-pressed SHS derived Ti_3AlC_2 powder obtained as a result of synthesis 1.1 TiAl + 2Ti + 2C.

Sintering temperature[°C]	Ti_2AlC	Ti_3AlC_2	TiC
20	9,1	74,3	16,7
1100	5,6	80,1	14,3
1200	-	90,8	9,2
1300	-	92,9	7,1

Table IV. Phase evolution of hot-pressed SHS derived Ti_3AlC_2 powder obtained as a result of synthesis $1.2Ti_3Al$ + Ti + C.

Sintering temperature[°C]	Ti_2AlC	Ti_3AlC_2	TiC
20	64,3	10	25,7
1100	57,5	21,1	21,4
1200	46,3	35,9	17,
1300	36,7	49,7	13,6

The example of XRD analysis of the sintered samples with highest amount of MAX phase is presented on Figure 2. Traces of corundum are contaminations left after the polishing process and were omitted in the phase analysis calculations.

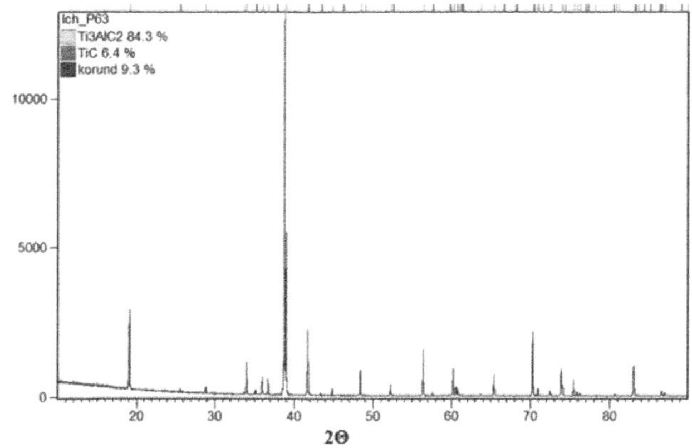

Figure 2. XRD pattern of the sample obtained as a result of hot-pressing of powder after the SHS synthesis of 1.1 TiAl + 2Ti + 2C at 1300°C .

The apparent densities measured for the samples hot-pressed at 1300°C are close to the theoretical densities of Ti_3AlC_2 and the differences are due to the presence of TiC impurities in the material. The values of apparent density are presented in the Table V.

Table V. Apparent densities of hot-pressed Ti_3AlC_2 materials [g/cm^3]

Reaction	1100°C	1200°C	1300°C
TiAl + 2Ti + 2C	3.03	3.72	4.30
1.1 TiAl + 2Ti + 2C	2.96	3.90	4.26
$1.2Ti_3Al$ + Ti + C	3.17	3.73	4.22

CONCLUSION

It can be stated that SHS is a suitable method for obtaining active powder precursors for hot-pressing of dense, polycrystalline materials in Ti-Al-C system, particularly Ti_3AlC_2. The upscaling process based on the previous researches showed that volume of reactants during SHS synthesis in some cases promotes formation of MAX phase material in the final product, thus powders with higher content (over 74 wt.%) of Ti_3AlC_2 and relatively low level of TiC impurities can be obtained. This method is not only possible but also effective and efficient.

As a result of these studies optimal conditions for hot-pressing of SHS derived active powder precursors were established. During the hot-pressing process further chemical reactions leading to elimination of impurities and formation of Ti_3AlC_2 are observed. It is worth to notice that especially elimination of titanium carbide from the system is very promising as it may affect wide range of properties of these materials. The increase of MAX phase volume was observed in case of all materials and the best result was achieved for powder being a product of 1.1 TiAl + 2Ti + 2C SHS synthesis hot pressed at 1300°C. In this case the amount of Ti_3AlC_2 phase was 92.9 wt%. Also investigations of relative densities shows, for the sample sintered at 1300°C, values close to the theoretical densities. Further researches on mechanical and functional properties of manufactured materials will be conducted.

ACKNOWLEDGMENTS

This work was supported by the National Science Centre under the grant no. 2013/11/B/ST5/02275

REFERENCES

[1] W. Jeitschko, H. Nowotny, F.Benesovsky, Kohlenstoffhaltige ternare Verbindungen (H-Phase). *Monatsh. Chem.* 94, 1963, p. 672-678

[2] H. Nowotny, Structurchemie Einiger Verbindungen der Ubergangsmetalle mit den Elementen C, Si, Ge, Sn. *Prog. Solid State Chem.* 2 1970, p 27

[3] M.W. Barsoum: The $M_{N+1}AX_N$ Phases a New Class of Solids; Thermodynamically Stable Nanolaminates- *Prog Solid St. Chem.* 28, 2000, pp. 201-281

[4] L. Chlubny, J. Lis "Properties of Hot-pressed Ti_2AlN Obtained by SHS Process", *Advanced Ceramic Coatings and Materials for Extreme Environments II*, 2012, Wiley, p. 171-177

[5] Y. Bai, X. He, C. Zhu, G. Chen " Microstructures, Electrical, Thermal, and Mechanical Properties of Bulk Ti_2AlC Synthesized by Self-Propagating High-Temperature Combustion Synthesis with Pseudo Hot Isostatic Pressing ", *J. Am. Ceram. Soc.*, 95 [1], 2012, p. 358–364

[6] X.H. Wang and Y.C. Zhou "Layered Machinable and Electrically Conductive Ti_2AlC and Ti_3AlC_2 Ceramics: a Review", *J. Mater. Sci. Technol.*, 26(5), 2010, p. 385-416

[7] M. Radovic, M. W. Barsoum; "MAX phases: Bridging the gap between metals and ceramics", *American Ceramic Society Bulletin*, 92 [3], 2013, p. 20-27

[8] Michel W. Barsoum; "MAX Phases: Properties of Machinable Ternary Carbides and Nitrides", Wiley, 2013, 436 pages

[9] J. Lis: Spiekalne proszki związków kowalencyjnych otrzymywane metodą Samorozwijającej się Syntezy Wysokotemperaturowej (SHS) - *Ceramics 44*: (1994) (*in Polish*)

[10] L. Chlubny, J. Lis; "Influence of Precursors Stoichiometry on SHS Synthesis of Ti_3AlC_2 Powders", *Processing and Properties of Advanced Ceramics and Composites V: Ceramic Transactions*, Vol. 240, 2013, p. 79-86

[11] A.G. Zhou, C.A. Wang, Z.B. Ge, L.F. Wu "Preparation of Ti_3AlC_2 and Ti_2AlC by Self propagating High temeprature Synthesis" *J. Mater. Sci. Lett.*, 21, 2001, p 1971

[12] M. Lopacinski, J. Puszynski, and J. Lis, "Synthesis of Ternary Titanium Aluminum Carbides Using Self-Propagating High-Temperature Synthesis Technique," *J. Am. Ceram. Soc.*, 84 [12], 2001, p. 3051–3053

[13] Y. Khoptiar, I. Gotman, "Ti_2AlC Ternary Carbide Synthesized by Thermal Explosion," *Mater. Lett.*, 57 [1], 2002, p. 72–76

[14] G. Liu, K.X. Chen, J.M. Guo, H.P. Zhou, J.M.F.Ferreira "Layered Growth of Ti_2AlC and Ti_3AlC_2 in Combustion Synthesis", *Mater. Lett.*, 61, 2007, p 779

[15] L. Chlubny, J. Lis, M.M. Bucko "Sintering and Hot-Pressing of Ti_2AlC Obtained By SHS Process", *Advances in Science and Technology*, 63, 2010, p. 282-286

[16] J. Lis, L. Chlubny, M. Lopacinski, L. Stobierski, M. M. Bucko " Ceramic Nanolaminates - Processing and Application", *Journal of the European Ceramic Society*, 28, 2008, p. 1009–1014

[17] L. Chlubny, M.M. Bucko, J. Lis "Intermetalics as a precursors in SHS synthesis of the materials in Ti-Al-C-N system" *Advances in Science and Technology*, 45, 2006, p. 1047-1051

[18] L. Chlubny, J. Lis, M. M. Bućko; "Titanium and aluminium based compounds as a precursors for SHS synthesis of Ti_2AlN", Strategic materials and computational design, eds. Waltraud M. Kriven, Yanchun Zhou, Miladin Radovic, Wiley, 2010, *Ceramic Engineering and Science Proceedings*, vol. 31, 10, p. 153–159

[19] L. Chlubny, J. Lis, M.M. Bucko: Preparation of Ti_3AlC_2 and Ti_2AlC powders by SHS method MS&T Pittsburgh 09: Material Science and Technology 2009, 2009, p. 2205-2213
[20] "Joint Commitee for Powder Diffraction Standards: International Center for Diffraction Data"
[21] H. M. Rietveld: “A profile refinement method for nuclear and magnetic structures.” J. Appl. Cryst. **2** (1969) p. 65-71
[22] L. Chlubny: New materials in Ti-Al-C-N system. - PhD Thesis. AGH-University of Science and Technology, Kraków 2006. (*in Polish*)

Author Index

Lightning Source UK Ltd.
Milton Keynes UK
UKOW06n1428150816

280738UK00001B/2/P

9 781119 236009